MOSQUITO SOLDIERS

MOSQUITO SOLDIERS

MALARIA, YELLOW FEVER, AND THE COURSE OF THE AMERICAN CIVIL WAR

ANDREW McILWAINE BELL

LOUISIANA STATE UNIVERSITY PRESS
BATON ROUGE

Published by Louisiana State University Press
Copyright © 2010 by Louisiana State University Press
All rights reserved
Manufactured in the United States of America
FIRST PRINTING

DESIGNER: *Amanda McDonald Scallan*
TYPEFACE: *text, Whitman; display, Mailart Rubberstamp*
PRINTER AND BINDER: *Thomson-Shore, Inc.*

Library of Congress Cataloging-in-Publication Data

Bell, Andrew McIlwaine, 1970–
  Mosquito soldiers : malaria, yellow fever, and the course of the American Civil War / Andrew
McIlwaine Bell.
    p. cm.
  Includes bibliographical references and index.
  ISBN 978-0-8071-3561-7 (cloth : alk. paper)
  1. United States—History—Civil War, 1861–1865—Health aspects. 2. Malaria—United States—
History—19th century. 3. Yellow fever—United States—History—19th century. 4. Mosquitoes as
carriers of disease—United States—History—19th century. 5. Soldiers—Diseases—United States—
History—19th century. 6. Soldiers—Diseases—Confederate States of America—History—19th
century. I. Title.

  E621.B54 2010
  973.7′75—dc22
                                                                    2009027446

For Dr. and Mrs. R. M. Bell, with filial love and gratitude

And the LORD said unto Moses, Yet will I bring one plague more upon Pharaoh, and upon Egypt; afterwards he will let you go hence: when he shall let you go, he shall surely thrust you out hence altogether.
—EXODUS 11:1

The subtle malaria of the rebel soil destroys and disables more Northern soldiers than all the wounds received from rebel arms.
—EDWIN C. BIDWELL, surgeon, Thirty-first Massachusetts Volunteer Infantry

CONTENTS

Illustrations follow page 71

PREFACE

Eyewitnesses to the American Civil War wrote a great deal about disease. The volume of information on malaria and yellow fever alone that can be gleaned from government publications, journal articles, personal accounts, and other sources is truly daunting (which should say something about their importance), and I have learned over the course of this project that it is impossible to describe every single clinical case of a particular disease when trying to produce readable epidemiological history. Therefore, in order to manage such a large amount of data, I have been forced to omit a number of anecdotes and case studies that, however fascinating, were not critical to my thesis. Readers interested in learning more about mosquito-borne illness during certain battles or campaigns are encouraged to consult my notes and bibliography as well as the bibliographies of texts referenced here.

As a medical/historical "realist," I believe modern science can help us better understand the medical problems of the past. Not every historian agrees. In fact, those who object to this approach sometimes accuse others who employ it of "presentism" or, worse yet, cultural smugness. But for me epidemiological history, like the other subbranches of the discipline, is a lot more lively and useful when we can compare what we know now with what our forebears knew in the past. The general rise in human life expectancy over the past two centuries (thanks in part to antiseptic surgery, antibiotics, and other groundbreaking developments) has buttressed the notion that immutable scientific laws—truths that transcend time and cultural traditions—exist within the field of medicine. My thesis, that mosquito-borne illness affected minds as well as bodies during the Civil War, rests on this basic premise. In other words, I believe that modern-day medicos know more about disease and the human body than nineteenth-century physicians did.

Of course, understanding antiquated medicine requires swallowing a healthy dose of historical context. Without it we put ourselves at risk of missing important philosophical questions and new ways of examining old practices. Bloodletting, for example, seen from the perspective of the time in which it was popular, probably cured some patients who wholeheartedly believed in its efficacy. Contemporary research on the mind's ability to heal the body shows us how such things could have been possible. But historical context can also trap scholars who overprescribe it, resulting in a sort of postmodern paralysis. A hypothetical (and thus politically expedient) example illustrates my point: a study of slavery which presented nothing but the misguided opinions of slaveholders would offend most modern readers. Historians are required to interpret data and make value judgments based on the prevailing wisdom of their times. For scholars exploring the annals of medicine, this means referencing modern medical concepts in addition to older ones. Doing so is also a way of paying tribute to those physicians who carefully compiled detailed clinical records so that future generations might be able to figure out what was killing their patients.

George Worthington Adams, Alfred Jay Bollet, H. H. Cunningham, Frank Freemon, Margaret Humphreys, and Paul Steiner know a great deal about what annihilated Civil War armies. I owe a debt of gratitude to these scholars—many of whom hold medical degrees—for making me comfortable with the idea of examining nineteenth-century medicine through modern scientific lenses in the first place. I encourage those interested in learning more about Union and Confederate field hospitals, surgical techniques, the health of black soldiers, the organization of the two medical departments, and diseases such as typhoid fever and dysentery to read their fascinating books.

A brief caveat for the casual reader is in order here: mosquito-borne illness was not the only reason Civil War campaigns and operations turned out the way they did. A host of other factors determined when, where, and how the North's and South's armies fought each other, and I have not tried to cover all of them in this specialized study. Readers desiring a comprehensive overview of the political and military events of the war should consult one or more of the brilliant studies written by Bruce Catton, Drew Faust, Gary Gallagher, James McPherson, Allan Nevins, James Robertson, Emory Thomas, and others too numerous to list here. My monograph is intended to complement, rather than compete with or replace, the works of these award-winning historians.

Those who prefer a quicker history lesson can scan the timeline of events in my appendix.

Producing scholarship is in some ways like fighting a war; one is forced to endure long periods of loneliness and boredom, followed by brief but terrifying attacks from faceless enemies (anyone who has ever been blasted in a double-blind review knows what I mean). Along the way certain people have made the enterprise more tolerable and as such deserve my thanks and praise. Tyler Anbinder, my dissertation director at the George Washington University, agreed to oversee this project when it was still in its larval stage. His comments, friendship, and encouragement over the years have been invaluable to me. Elizabeth Fenn, another GW professor (now at Duke), served as an academic gadfly and stung me into action with a deft response to my suggestion that she write a book on malaria during the American Revolution. She was also kind enough to share her notes and wisdom on the pitfalls and pleasures of pursuing epidemiological history. Dale C. Smith, a talented medical historian working at the Uniformed Services University of the Health Sciences in Bethesda, Maryland, took time out of his busy schedule to read drafts and introduce me to the murky world of nineteenth-century medicine—thank you, Dr. Smith. Stephen Greenberg and his staff at the History of Medicine Division of the National Library of Medicine are truly wonderful human beings. They pulled books off shelves, suggested sources I had not even considered, and greeted me with a smile every time I entered their domain. Terry Reimer, Kari Turner, and others at the National Museum of Civil War Medicine in Frederick, Maryland, were equally friendly and allowed me to comb through their collections for hours on end. Two Georgetown professors, Peter Armbruster (biology) and John McNeil (history), gave me advice and encouragement early on in my research. Janie Morris and David Strader helped me navigate Duke's special collections library. Beverly Tetterton at the New Hanover County Public Library in North Carolina; Casey Greene at the Rosenberg Library in Galveston, Texas; Philip Montgomery at Rice University; and Jacob Crawford at the National Institutes of Health contributed their time and ideas. Dee Dee King, Genevieve Croft, and the unsung heroes at the Library of Congress and National Archives also deserve thanks. So does Stephen Bryce, the GIS genius responsible for the maps presented in this study. Richard Sommers and Michael Knapp at the U.S. Army Military History Institute in Carlisle, Pennsylvania, provided funding and opened their files for my perusal. My fellow doctoral students at the George Washington University—especially Craig, Varad, and Kata (and her better half, Raul)—offered support and valuable feedback.

And last, but certainly not least, are the hundreds of thousands of soldiers who died from disease during the Civil War. Like the country as a whole, I owe them a debt I can never repay. Although their less-than-glorious cause of death meant that their names were never chiseled in stone on battlefield monuments, they too gave what Abraham Lincoln described as "the last full measure of devotion." Thank you, gentlemen—the publication of this book proves that you have not been forgotten.

Mosquito Soldiers

Introduction

MANY OF US can still remember the sense of awe we experienced the day we learned that 620,000 Americans had died during the Civil War. It may have been a high school history teacher who first mentioned the number to us, or perhaps we picked it up while reading one of the myriad military histories that have been published in the century and a half since the conflict ended. In any case it was clear that from that moment forward our understanding of this pivotal moment in our country's past would never be the same. The number dwarfed every other war-related statistic we had encountered, making us wonder how our predecessors could have slaughtered one another at such an appalling rate in only four years' time, especially since every other military conflict we had learned about, including the bloodbath in Europe in the mid-twentieth century, produced fewer American casualties. What was it like to be part of such a violent struggle? Our minds instinctively conjured up scenes of bayonet charges, exploding shells, and battlefields littered with corpses clad in soiled blue and gray uniforms, the same images that have been promoted by media moguls and well-intentioned instructors since the nineteenth century. But few of us were ever told that another story exists apart from the flanking maneuvers and combat heroics that have long preoccupied scholars, this one even messier, more tragic and profane, and more nauseatingly human than anything we might have imagined. For in truth the overwhelming majority of the 620,000 men who perished during the Civil War did not die from gunshot wounds or saber cuts but were victims of dangerous maladies that frequently made life for earlier generations of Americans gruesome and short.

It is indeed startling to realize that troops on both sides were more than twice as likely to be killed by deadly microorganisms as enemy fire. In the U.S. Army alone disease accounted for 66 percent of total wartime fatalities; Confederate numbers were comparable. And of the various maladies that plagued both armies, few were more pervasive than malaria. Between 1861 and 1866 over a million white soldiers serving in the Union army were diagnosed with the disease, along with 152,141 of their black comrades in arms. The ailment was a problem for northern troops in every Union military department (repeat infections pushed the average annual incidence rate beyond 100 percent in some areas) and was responsible for every seventh case of sickness among Confederate soldiers stationed east of the Mississippi River in 1861 and 1862. It also created medical complications for men who were suffering from one or more of the various other diseases that plagued Civil War armies, including chronic diarrhea, the biggest pathologic killer of the war. After Appomattox Federal medical officials correctly concluded that northern soldiers suffering with diarrhea and malaria at the same time had succumbed to death more readily than those who were suffering from a gastrointestinal ailment alone. Dysentery (frequently characterized by blood in the stool) and diarrhea together were responsible for more than 40,000 Union deaths.[1]

Because unhealthy armies inevitably affect military planning, it should come as no surprise to learn that the ubiquity of malarial infections among northern and southern troops helped shape the events of the Civil War. The disease was one of the main contributors to the South's annual "sickly season"—the time of year, usually in the summer and fall months, when the region's insects, parasites, and bacteria were most active and made life unbearable for all but the hardiest of souls. Skittish about risking the health of their troops, military commanders, especially northern ones, routinely avoided operations during this period. For many of them the season was similar to winter, the time of year when armies traditionally withdrew from active campaigning to await the return of better weather.

Malaria and the sickly season the disease helped create offered the South advantages over its enemies. Primarily fighting a defensive war, southern military planners realized that their Union counterparts were unlikely to launch major offensives in certain areas of the Confederacy during the unhealthiest months of the year. Armed with this knowledge, they transferred troops from sickly districts to other more threatened fronts or reinforced armies tasked with taking the war into the northern states.

But malaria was not the only ailment that contributed to the South's reputation for dangerous diseases. Yellow fever outbreaks in Charleston, Mobile, New Orleans, Norfolk, Savannah, and other southern cities during the 1850s killed thousands of people and shaped outsiders' perceptions of the region as an unhealthy place to live and do business. When war came, southerners expressed hope that the dreaded disease would aid their cause, and occasionally their wishes were fulfilled. Actual cases of yellow fever as well as the trepidation the malady caused among northern and southern commanders (equally important from a military standpoint) affected operations in a number of the South's coastal areas. In the wake of the Confederacy's collapse, however, federal officials largely dismissed the importance of these incidents and the urgent concerns they had produced during the war. In his magnum opus, *The Medical and Surgical History of the War of the Rebellion* (prepared under the supervision of the surgeon general's office and first published in the 1870s), Joseph Janvier Woodward downplayed the significance of yellow fever. Gathering reports from army surgeons stationed in various departments across the South, Woodward determined that there had been only 1,355 cases of yellow fever and 436 deaths from the disease among Union troops.

Yet these figures do not include the hundreds of Union seamen, southern civilians, and Confederate soldiers who also contracted the virus while the fighting lasted. Nor do they reveal the psychological impact the disease had on nervous officers who were assigned to areas with a history of epidemics and were justifiably worried about their health and that of the men under their command. For them yellow fever was a terrifying and mysterious southern disease that could appear out of thin air at any moment and wipe out an unacclimated brigade, division, or crew in a matter of weeks. They had no way of knowing—as the military-medical establishment did later, with the benefit of hindsight—that yellow fever would kill fewer men during the war than gastrointestinal ailments or typhoid fever. The disease struck fear into the hearts of the men and women who witnessed America's sectional conflict because they knew all too well that it had annihilated huge numbers of people living in southern cities during the first half of the nineteenth century. Epidemics flared up and then ended almost as quickly as they began, leaving behind ghastly scenes of death and devastation. In order to subdue the South, Union soldiers would need to occupy these sites. Confederates were delighted that their enemies would be battling "yellow jack" as well as southern armies, but they quickly realized that the disease could work against their own forces as well. Outbreaks appeared during the war which snuffed out the lives of Rebel soldiers and civilians alike. Yellow fever, like malaria, was a threat primarily during the summer and autumn months.

Unbeknownst to the Civil War generation, both of these diseases were being transmitted by the swarms of mosquitoes that inhabited nearly every region of the South in the nineteenth century. Soldiers on both sides frequently complained about these annoying insects that fed on their blood, buzzed in their ears, invaded their tents, and generally contributed to the misery of army life. Little did they suspect that these pests were also helping to shape the larger political and military events of the era. Throughout the war the South's large mosquito population operated as a sort of mercenary force, a third army, one that could work for or against either side depending on the circumstances. It is for this reason, and because the average enlisted man in the Civil War encountered the insects almost everywhere he went, that this work is entitled "Mosquito Soldiers."

Medicine in the 1860s was a hodgepodge of ancient shibboleths, folk remedies, and sound scientific practices. As inheritors of the Greek tradition, Civil War–era physicians viewed the body as a precariously balanced concoction of chemicals which any number of external and environmental stimuli could alter. The key to preserving a patient's health lay in maintaining his or her internal equilibrium. Drugs and therapies that induced vomiting, sweating, or the evacuation of the bowels or bladder were thought to assist in bringing a diseased body back into balance. Because patients and their families insisted on some sort of treatment for serious medical conditions, most "regular" doctors were willing to prescribe chemicals and harsh therapies that we now know are harmful to the human body. Modern audiences can be forgiven for cringing at the crudest nineteenth-century therapeutics (dosing a patient with mercury, for example, seems cruel by today's standards). But we should avoid condemning the physicians who employed them, especially since these treatments proved effective from time to time. Understanding why requires an appreciation of the role faith plays in promoting healing. For those who lived before microbiology became an established science, doing something for a patient was usually better than doing nothing at all, especially if a person's life were at stake. If a sick person pulled through after ingesting a certain compound or chemical, the attending physician and his observers would naturally credit the patient's recovery to the prescribed therapy, which ensured its continued use within the community. Medicine, like politics, was a mostly local affair in the 1800s.[2]

Trial-and-error therapeutics of this sort arose in part because of the poor quality of medical education available at the time. Of the eighty or so medical schools that were open on the eve of the Civil War, most were run as for-profit ventures rather than cutting-edge research centers. Anyone willing to pay school administrative fees and purchase lecture "tickets" from professors could earn a medical degree. Students sat through the same lectures two years in a row (administrators believed that this approach would allow the information to sink in better) before apprenticing themselves to a practicing physician. Schools did not regulate these apprenticeships or even require that their students undergo supervised clinical training.[3]

Government oversight was equally lax. The leveling spirit of the Jacksonian era convinced a number of Democratic state legislatures that medical licensing laws were elitist conspiracies designed to perpetuate aristocratic privilege. The repeal of these laws in the 1830s and 1840s allowed the profession to slide even further into chaos. By the time the Civil War began, nearly any person who wanted to call himself or herself a doctor could hang out a shingle and treat patients.[4]

But within this chaotic climate there were also clear signs of scientific progress. Practitioners at Massachusetts General Hospital discovered the benefits of anesthesia in 1846, which allowed Western physicians to perform humane surgery for the first time. In Paris Americans studying under Pierre Louis were learning to place a greater emphasis on natural healing and to question the usefulness of therapeutics that had not been subjected to close clinical scrutiny. This greater reliance on empirical observation led to changed perceptions of disease. By the middle of the nineteenth century the most enlightened practitioners "began to think more in terms of discrete disease entities and disease-specific causation and less in terms of general destabilizing forces that unbalanced the body's natural equilibrium."[5]

With no understanding of bacteriology or virology, however, physicians theorized that certain diseases were caused by invisible poisons that floated through the air. Malarial fevers were blamed on "miasmas" produced by decomposing organic material (especially in and around swamps), while yellow fever was attributed to a mysterious filth that attached itself to certain types of clothing and traveled aboard ships. The most potent weapon Civil War surgeons had in their fight against malaria was quinine. The advantage this drug gave to Union forces cannot be overstated (in fact, one could argue without too much hyperbole that a more appropriate subtitle for this book might have been "How Quinine Saved the North"). Yankee surgeons doled out more than nineteen tons of the medicine and used it as both a specific and prophylactic among troops stationed in the sickliest regions of the South. The Confederacy, on the other hand, experienced quinine shortages for most of the war, which meant malarial fevers among Rebels went unchecked more often than not. Southern civilians also suffered. Those who believed that life in their part of the country depended on the drug were less inclined to support a government that could not supply it. The paucity of quinine was one more inconvenience imposed upon a southern public that was already frustrated over impressment, enemy raids, conscription, and bread shortages.[6]

Of course, nineteenth-century southerners were not the only people in history ever to have endured war and pestilence at the same time. In fact, epidemic diseases seem to flourish whenever human beings come together to slaughter one another in the name of an agenda or set of beliefs. Thirty years ago William McNeill described how smallpox had allowed a handful of Spanish conquistadors to subjugate an entire continent, and more recently, Elizabeth Fenn has shown that this disease also influenced the events of the American Revolution. For whatever reason Civil War historians have shied away from similar studies even while acknowledging that far more soldiers died in sickbeds than on battlefields.[7]

This lack of scholarly attention has produced one of the few remaining gaps in the historiography of the war. More studies are needed which explain how specific diseases affected military operations and strategy during the conflict. Few other ailments sickened as many troops and civilians as malaria, and no other disease (except perhaps cholera) produced more public panic throughout the nineteenth century than yellow fever. To date most medical histories of the conflict have given short shrift to both illnesses. As a result, historians are left

with an inchoate picture of daily life in the wartime South and an
incomplete understanding of why various campaigns turned out the
way they did.

The importance of this study, then, is threefold; first, it reinterprets familiar events from an epidemiological perspective and adds to a medical historiography that has long been in need of a transfusion of scholarly interest. Even though disease-related fatalities outnumbered combat deaths two to one, far more intellectual energy has been expended on dissecting individual battles than in researching what one prominent Civil War historian fittingly referred to as "the grimmest reaper." It is the author's hope that this work will encourage other students of the period to investigate the various other diseases that beset both soldiers and civilians and uncover the causal links that exist between human health and history.[8]

Second, malaria and yellow fever made a difference in the outcome of several Civil War campaigns. These diseases not only sickened thousands of Union and Confederate soldiers but also affected the timing and success of certain key military operations. Some commanders took seriously the threat posed by the southern disease environment and planned accordingly; others reacted only after large numbers of their men had already fallen ill. Simply stated, had mosquito-borne illness not been part of the South's landscape in the 1860s, the story of the war would be different.

Finally, by focusing on two specific diseases, which share a similar vector, rather than a broad array of Civil War medical topics, this study provides readers with a clearer understanding of how environmental factors serve as agents of change in history. Regrettably, our arrogance has deceived us into believing that our actions alone determine what happens in the course of human events. But in truth *Homo sapiens* inhabit a broader natural world built on a series of complex, interdependent relationships that, when altered, can produce catastrophic results. War represents not only a breakdown of human social and political relations but also the disintegration of the existing environmental order. The large armies required by both the Confederate and Union governments accelerated the development of diseases that thrive on human hosts, which in turn affected the ability of these forces to carry out the instructions they received from military commanders.

Although malaria and yellow fever were only two of a multitude of maladies that afflicted Union and Confederate troops, few diseases were greater agents of change. These illnesses not only affected military planning and influenced medical practices on both sides but also helped change the lives of nearly every American who witnessed the war. Shortages of malarial medicine transformed the ideal southern woman of leisure into a black market smuggler and made plantation life increasingly arduous. Infections among African-American troops weakened pseudoscientific claims that for generations had served as convenient justifications for slavery. Southern urbanites learned the value of rigorous sanitation and quarantine practices during the Union occupation and then endured the horror of new yellow fever outbreaks once it ended. Federal soldiers were infected with malarial parasites that had largely disappeared from their home communities by the 1860s and reintroduced them into non-immune northern areas after the war. Confederate quartermasters watched helplessly as yellow fever plagued important port cities, disrupting critical supply chains and creating public panics. All of these changes were wrought by a tiny insect whose omnipresence in the United States seemed perfectly natural and nonthreatening to previous generations.

In 1861 the coasts and prairies, woodlands and wetlands, and bayous and bogs of the newly formed Confederacy were teeming with mosquitoes that would eventually stymie military campaigns, kill thousands of soldiers and sailors, and cause pain and suffering for countless southern civilians. The important role these insects played cannot be ignored by any scholar aspiring to understand the Civil War in all its wonderful and dizzying complexity.

AEDES, ANOPHELES, AND THE SCOURGES OF THE SOUTH

THE SUMMER OF 1835 was a stressful time for Mrs. Zachary Taylor. In June she learned her daughter Sarah Knox had married a dashing young army lieutenant named Jefferson Davis and dutifully followed her new husband from Kentucky to his virgin estate on the Mississippi River below Vicksburg. Although Colonel Taylor expressed fatherly doubts about the wisdom of letting "Knoxie" become a soldier's wife, Mrs. Taylor was less concerned with her daughter's choice of a spouse than with the dangerous diseases she knew lurked in the shadowy, stagnant bogs that surrounded Davis's "Brierfield" plantation. Fifteen years earlier she and her four children, including Sarah, had fallen ill while stationed with her husband's regiment in the swamps of Louisiana. The two youngest girls did not survive, and Mrs. Taylor herself had nearly died. The thought of losing another child to the mysterious and unhealthy climate of the Deep South was more than she could bear.

Sarah tried to assuage her mother's fears. "Do not make yourself uneasy about me," she wrote in August. "The country is quite healthy." The following month, however, both she and her husband were stricken by a severe illness while visiting relatives in West Feliciana, Louisiana. On the fifteenth of September, at the height of their agonizing ordeal, Davis staggered out of his sickbed to comfort his ailing wife, but Sarah, delirious with fever, failed to recognize him. Instead, in a fit of madness she sang a popular nineteenth-century song called "Fairy

Bells" before closing her eyes forever. Davis never got a chance to say good-bye to his twenty-one-year-old bride and was forced to mourn her loss while still very ill. The future president of the Confederacy grieved in seclusion at Brierfield for the next eight years.[1]

Like most Americans of her era, Mrs. Taylor would have likely attributed her daughter's tragic death to the poisonous air that was said to pervade unhealthy areas of the country such as the Mississippi Delta. Conventional wisdom since colonial times held that strange and virulent vapors continually wafted through the atmosphere of the warmer regions of North America and created health problems for anyone unfortunate enough to breathe them in, especially those born in healthier climates. What exactly caused the air to turn lethal, however, was still in dispute among physicians for most of the nineteenth century. Endless theories circulated in medical journals or were discussed at conventions at a time when medicine was more art than science. Some practitioners, perhaps a majority, believed decomposing animals and plants produced the noxious "miasmas" that sickened their patients. Others thought electrical charges in the ozone were the culprit. Still others rejected the "bad air" theory altogether and instead blamed excess hydrocarbons in the blood.[2]

In reality Sarah Taylor Davis had died of an insect bite. One night in late August or early September a female mosquito carrying a dangerous strain of malaria surreptitiously sliced through her skin, sucked up her red corpuscles through its straw-like proboscis, and unwittingly released into her bloodstream a dozen or so malarial sporozoites that it had picked up from a previous victim. Within minutes these sporozoites found their way into Sarah's liver, where they transformed over the next two weeks into schizonts, each containing thousands of smaller organisms called merozoites. When the mature schizonts eventually burst, the merozoites poured out like tiny soldiers and invaded her red cells in order to reach the next stage of their development. As these parasites rapidly multiplied, dead and dying corpuscles clung to the walls of Sarah's capillaries and healthy cells, creating a dam that blocked the flow of blood to her vital organs.[3]

Malaria is a parasite transmitted by *Anopheles* mosquitoes, a genus that prefers to breed in stagnant, sunlit pools of fresh water and can be found in most regions of the country. The adult female requires a blood meal to ovulate and can lay between one and three hundred eggs at a time. Symptoms of malaria include chills, shakes, nausea, headache, an enlarged spleen, and a fever that spikes every one to three days depending on the type of malaria and its parasitic cycle. In all likelihood Sarah Davis was killed by *Plasmodium falciparum*, one of four types of malaria that infect human beings. Of the other three— *vivax*, *malariae*, and *ovale*—only *Plasmodium vivax* was once common in the United States. *Vivax* alone rarely proved fatal to its victims, but *Plasmodium falciparum* was often deadly.[4]

Nineteenth-century physicians categorized malaria according to how often these fever spikes, or "paroxysms," occurred. A "quotidian" fever appeared once every twenty-four hours, a "tertian" every forty-eight, and a "quartan" every seventy-two. *Plasmodium vivax* was commonly referred to as "intermittent fever," "ague," "dumb ague," or "chill-fever," while *Plasmodium falciparum* was known as "congestive fever," "malignant fever," or "pernicious malaria" because of its lethal effect.[5]

Patients diagnosed with "remittent fever" experienced febrile symptoms that, as the name suggests, periodically went into remission. But they did not disappear entirely (and temporarily) in the same way that so-called intermittent symptoms did. Like most nineteenth-century descriptions of disease, the term *remittent* was somewhat nebulous. Yet a handful of studies conducted in the 1830s and 1840s suggest that many of these fevers were the result of repeat plasmodial infections.[6]

Sarah Davis was one of countless Americans who contracted malaria during the nineteenth century. As settlers cleared virgin forests from Savannah to St. Louis to make way for the cotton and wheat farms that drove the antebellum economy, they inadvertently created a plethora of new breeding sites for anopheles. At a time when mosquito-control measures such as spraying the insecticide chlorophenothane (DDT) were still unknown, clouds of insects swarmed wagon trains and slave coffles, sparking complaints about "fever" and "ague" in the South and West, where crude housing, poor drainage, and regular flooding aided in the spread of the disease. Daniel Brush's experience on the frontier was typical. In 1820 he and his family moved from Vermont to southern Illinois in search of better economic opportunities and wound up with a handful of other Yankee families in a small settlement called "Bluffdale," four miles east of the Illinois River. When the entire community came down with malaria during the first harvest, Brush recorded his fellow settlers' suffering: "Many had the real 'shakes' and when the fit was fully on shook so violently that they could not hold a glass of water with which to check the consuming thirst that constantly beset them while the rigor lasted, nearly freezing the victim." He went on to describe the fevers that followed these fits as putting "the blood seemingly at boiling heat and the flesh roasting."[7]

Other observers noticed the prevalence of the disease in the West. During a tour of the United States in the 1840s, the English author Charles Dickens encountered so many "hollow-cheeked and pale" malaria victims that he forever remembered the region where the Mississippi and Ohio rivers converge as "a breeding-place of fever, ague, and death." St. Louis also seemed unhealthy to Dickens, despite its residents' claims to the contrary. One Illinois physician's frequent contact with malaria convinced him that he could accurately diagnose patients just by learning where they lived, while another practitioner thought the malaria victims he saw, even children, looked "prematurely old and wrinkled." Country doctors from all over the Northwest published articles in medical journals on the best ways to identify and combat the mysterious disease that plagued their communities.[8]

But while malaria made life difficult for Westerners, it made life nearly intolerable for southerners at certain times of the year. *Plasmodium falciparum* occurred almost exclusively below the thirty-fifth parallel and was especially problematic in the states of the Deep South such as South Carolina and Georgia. Short, mild southern winters substantially lengthened the breeding season for anopheline mosquitoes and made *Plasmodium vivax* infections as common as colds in some areas. White southerners dreaded the annual arrival of the "fever season" (which lasted for a variable length of time between late spring and early autumn depending on the location), and those who could afford it escaped to seaside cottages or fled northward in search of healthier climates. During a visit to lowland South Carolina in the 1850s, landscape architect Frederick Law Olmstead noticed that the overseer on one plantation moved inland to higher ground during the "sickly season" (outside the flight range of anopheles) to escape the "swamps" and "rice-fields" that made life at night "dangerous for any but negroes." The widespread belief among whites that blacks were immune to malaria, which served as a convenient justification for slavery, had some basis in scientific truth. West Africans inhabited malarial environments for thousands of years before being brought to America and developed a degree of genetic resistance which they passed on to

their offspring. But by the mid-nineteenth century Africans from all
over the subcontinent were intermixing with one another as well as
with Indians and Europeans, which meant that many blacks were also
susceptible to malaria.[9]

Although neither blacks nor whites understood what caused malaria, they both agreed that the farther south one traveled, the more
prevalent the disease became. Slaves from states such as Virginia and
Maryland feared being sold to planters in the lower South in part because of the malarial poison they knew plagued the area. John Greenleaf Whittier's poem "The Farewell" captures the concern many black
families must have felt for the health of their loved ones who were
forced to move farther south:

> Gone, gone, sold and gone
> To the rice swamp dank and lone,
> Where the slave-whip ceaseless swings,
> Where the noisome insect stings,
> Where the fever-demon strews
> Poison with the falling dews,
> Where the sickly sunbeams glare
> Through the hot and misty air:—[10]

Whites also worried about the endemic malarial fevers of the
South. Residents of lower Louisiana and Mississippi who lived in
swamps near the Mississippi River routinely evacuated their homes
to escape the strange sickness that mysteriously appeared whenever
it flooded. Together with portions of western Tennessee, these areas
were considered the most malarious in the Mississippi River Valley.
Other locations infested with mosquitoes, such as the algae-covered
bogs of south Alabama, coastal Carolina, and the Florida panhandle,
were thought to be uninhabitable or exceedingly dangerous places for
whites during the summer months. A physician practicing in tidewater
North Carolina believed the prevalence of malaria in his neighborhood
"would appear incredible to those whose experience has been confined
to more healthy localities." Indeed, military personnel stationed in the
South for the first time were horrified by the unhealthiness of the
climate. An army physician from St. Louis assigned to a fort on the
Suwannee River in Florida saw so many cases of intermittent fever
that he was convinced that even the pets in the place suffered from the

disease. He and two-thirds of the soldiers under his care fell ill while on duty. Military doctors serving in Baton Rouge, Louisiana, considered one of the sickliest posts in the United States in the early 1800s, saw an average of 824 cases of malaria per 1,000 troops each year. Fort Gibson in Arkansas and Fort Scott in Georgia were also reputed to be malarial pestholes. From the burgeoning cotton estates of the Texas plains to the well-established tobacco and rice plantations of the East, anopheles mosquitoes fed on black and white bodies indiscriminately, transmitting plasmodium parasites and sickening their human prey in the process.[11]

The prevalence and severity of malaria in the South helped shape both northerners' and southerners' views of the region. For northerners this view was mostly negative. By the mid-nineteenth century long winters and infrastructure improvement projects spearheaded by industrious free laborers had all but eliminated malaria from New England and given rise to a generation of Northeasterners who had never been exposed to the disease. Like many Americans today, antebellum New Englanders viewed malaria as a "tropical" malady that was only a problem for people who lived in the unsanitary and underdeveloped regions of the world, which included the southern United States. Southerners' repeated warnings about the health risks their local fevers posed to outsiders reinforced this negative image. So too did the stories shared by returning sailors about the intermittent fevers they had contracted while visiting cities such as Mobile and New Orleans. Malaria even helped convince northern insurance firms of the need to charge higher premiums to their southern clients and hindered the North-to-South flow of investment capital and people. In contrast, some southerners adopted a sort of perverse pride in the ailments they considered their own. Repeated infections with the same strains of malaria over a period of years meant southerners who stayed in one place became accustomed to the fevers in their neighborhood and developed a limited immunity against them. A number of southern physicians practicing during the antebellum period even went so far as to call for the creation of separate medical schools and publications to allow southern students the chance to learn more about the diseases that were indigenous to their region.[12]

Malaria, however, was not the only mosquito-borne illness that contributed to the image of the South as an unhealthy place to live. Yellow fever, a disease once found as far north as New York and Philadelphia, became a uniquely southern problem by the middle of the nineteenth century. The reasons why it became confined to the South remain unclear, although the North's longer winters, quarantine and sanitation practices, and trade patterns may have played a role. Outbreaks in Charleston, Galveston, Mobile, New Orleans, Norfolk, Savannah, and other southern cities killed tens of thousands of people and created a level of physical and emotional suffering that can scarcely be imagined today. Victims in the advanced stages of the disease bled from the nose and mouth; suffered excruciating headaches, fever, and jaundice; and, worst of all, vomited a substance resembling coffee grounds (half-digested blood) caused by internal hemorrhaging, which was a telltale sign of the virus. Fatality rates during epidemics ranged from 15 to over 50 percent, but those who survived acquired lifetime immunity. Nineteenth-century Americans lived in fear of the disease they called the "scourge of the South," and public panics often followed the first sign of an outbreak.[13]

Unbeknownst to antebellum physicians, yellow fever was being transmitted from person to person by the *Aedes aegypti* mosquito, a species that inhabits the southern United States and lays its eggs in hollow logs and artificial receptacles containing freshwater. Filthy southern cities offered nearly limitless incubation pools for *Aedes* eggs in the form of horse troughs, barrels, clogged gutters, and trash in the streets filled with rainwater. Winter frosts limited the activity of the mosquito and prevented yellow fever from ever becoming endemic in North America, but the virus was continually reintroduced by cargo ships arriving from the Caribbean, where it existed year-round. A single infected sailor or mosquito on board could spread yellow fever to the local *Aedes aegypti* population and cause widespread misery and death.[14]

Just such a scenario unfolded in New Orleans in the summer of 1853 in an outbreak that claimed over eight thousand lives in a matter of weeks. Panic-stricken residents closed their shops and businesses and fled in all directions to escape the poisonous atmosphere of the city, leaving behind unburied corpses that swelled and burst in the blistering Louisiana heat. One man was in such a hurry to escape that he decided, in the interest of saving time, not to press charges against a criminal who had stabbed him. New Orleans's Charity Hospital was soon filled to capacity, and city officials burned tar on the street corners and fired cannons each morning and evening in an attempt to purify the supposedly unclean air. Priests and ministers were "called upon every hour of the day and night" to comfort the dying and their relatives, while evacuees spread the virus to other towns along the Gulf Coast and Mississippi River, killing thousands more. Four more major yellow fever epidemics over the next five years helped the Crescent City earn a reputation as a den of despondency and death.[15]

A year after yellow fever razed New Orleans, it appeared in Savannah, Georgia, and Charleston, South Carolina. Obituaries printed in the Savannah papers show a disproportionate number of the fatalities were German and Irish immigrants who had never been exposed to the disease and therefore had no immunity against it. Scores of men and women with last names like Hanns, Krauss, McDonald, and O'Donnell who had come to Georgia in search of opportunity found an early grave instead. The disease sickened Savannah's bakers and kept farmers in the surrounding countryside from bringing their produce to market, resulting in food shortages, which compounded victims' suffering. All one desperate resident could find to eat was "three or four spoiled hams." Hungry, sick families were left vulnerable to looters, who took advantage of the chaos and began to plunder private homes.[16]

In Charleston the virus also targeted Irish and German immigrants in addition to nonimmune Americans from other parts of the United States whom native South Carolinians considered part of the "foreign population." But by the end of September it "invaded all parts of the city" and afflicted residents of every "race" and "age," including a number of wealthy locals who, ironically enough, had postponed their usual summer travel plans to avoid the cholera outbreaks that were said to be raging in the northern states at the same time. The Charleston papers printed the daily death toll (nineteen persons died in one twenty-four-hour period), and the city's mayor enacted a strict quarantine on all incoming ships. People living in cities surrounding Charleston and Savannah panicked. When yellow fever appeared in Augusta, Georgia, in mid-September, a stampede of "four or five hundred persons" fled the city. Officials in Wilmington, North Carolina, were so unnerved by the reports coming out of Charleston and elsewhere that they imposed a one-hundred-dollar-a-day fine on anyone found inside their city who was known to have come in contact with the disease. By the time winter arrived and halted the epidemics in Charleston and Savannah, more than sixteen hundred people were dead.[17]

Yellow fever plagued the Gulf Coast of Texas as well. Epidemics hit Galveston and Houston throughout the 1830s, 1840s, and 1850s and periodically spread along the railroad lines to surrounding towns such as Cypress, Hempstead, and Montgomery. The suffering caused by these outbreaks was horrific. In 1839 Ashbel Smith was practicing medicine in Galveston when the city was struck with yellow fever. During one house call he encountered a twenty-five-year-old man he referred to as "L" whose "condition was truly awful." Clad only in a shirt, L was delirious with fever, bleeding from his mouth, and "lying on the bed in every variety of position, sometimes on his face and knees." Vomit and excrement were everywhere. Another of Smith's patients, a baker named Mat, was "ejecting black vomit freely" and babbling incoherently when the doctor arrived. The unfortunate man died within a matter of hours. Such gruesome stories of misery and death reached outsiders and added to the South's reputation as a bastion of insalubrity.[18]

An outbreak that struck Norfolk, Virginia, in 1855 provides a useful glance at how yellow fever could be introduced into a community. In June the steamship *Ben Franklin* arrived from St. Thomas, where yellow fever was known to be raging, and was immediately directed to the city's quarantine station. After twelve days and with no reports of sickness from the ship's captain, Norfolk's health officer cleared the *Franklin* and allowed it to steam into the harbor for repairs. In early July a mechanic died of yellow fever after working in the ship's engine room. In all likelihood he was bitten by an infected *Aedes aegypti* that had flown aboard the *Franklin* in St. Thomas, attracted by the sweaty bodies and pools of stagnant water in the vessel's hull. Soon after the repairman's death, the mosquitoes of Norfolk and nearby Portsmouth created the same ghastly scenes of human suffering which had horrified the citizens of Charleston, New Orleans, and Savannah in the previous two years. Coffins piled up in local graveyards, and terrified crowds flocked to the countryside to escape the disease. One Presbyterian minister in Norfolk remembered the swarms of flies that "collected about the doors and windows" outside the houses of the dying. Nearly two thousand Virginians perished during the epidemic.[19]

Northerners reacted to these macabre episodes with a mixture of compassion and loathing. Relief committees in northern cities raised money for victims, while nurses, doctors, and druggists headed south to give aid to the sick, occasionally contracting the disease and dying in the process. But other northerners were less sympathetic and interpreted the outbreaks as proof of the inferiority of the southern way of life. A Unitarian minister in Washington blamed Portsmouth's "evil institutions" and "unwise laws" for the filthy conditions that allowed yellow fever to develop there. Another somewhat sanctimonious New York preacher told his congregation that one blast of cold northern air would mean more to the yellow fever sufferers in Norfolk than "a thousand physicians and surgeons convened there." And a number of abolitionists believed the disease was divine punishment for the sin of slavery and argued (correctly, it turns out) that the malady was a consequence of the slave trade.[20]

At a time when "Bleeding Kansas" and the Fugitive Slave Law were fueling sectional tensions, southerners reacted defensively to these condemnations but were at a loss about how to control the disease that had grown to be identified with their section of the country. The medical community offered few answers and could not agree on the origin of yellow fever (either locally or outside the country) or even whether it was contagious or not. As a result, quarantines and sanitation measures, policies that in all likelihood helped eliminate yellow fever in northern cities in the early nineteenth century, were unevenly enforced in the South. City officials and newspaper editors often downplayed or denied the presence of the disease until the last minute in order to stave off public panics and the economic downturns that inevitably followed outbreaks. Businessmen who relied on maritime trade for a living exacerbated the problem by lobbying against lengthy quarantines that might cut into their profits.[21]

While Southern urbanites dreaded the return of yellow fever to their cities, they also knew that newcomers and visitors, people born in areas free from the disease, were especially vulnerable to its ravages. Locals who lived in endemic regions and contracted mild cases as children or adults enjoyed lifelong immunity. Many blacks seemed immune to the disease in the same way they did with malaria. Predictably, Southern whites observed these differences and cited them in proslavery screeds.[22]

In contrast, most of the New England farm boys and shopkeepers who volunteered to invade the South in 1861 had never been exposed to either yellow fever or malaria. And while their western comrades-in-arms had long been familiar with "ague," they too were susceptible to the unique and potentially dangerous strains of the disease that flourished below the Mason-Dixon Line. In short the outbreak of the

Civil War gave the arbovirus known to the Spanish as "negro vomito" and the multiple malarial parasites of the South, including *Plasmodium falciparum,* access to a legion of new, nonimmune hosts. An army of mosquitoes awaited the men in blue which was every bit as lethal as the gray-clad army they would face on the battlefield. And before the war ended, this army slew thousands of Union troops and influenced the strategic and operational thinking of their commanders.

The same concerns Mrs. Taylor had expressed to her daughter in the summer of 1835 were on the minds of the top brass in Washington in the spring of southern secession. Two months before the Battle of Bull Run, Commanding General Winfield Scott warned Major General George McClellan that the biggest threat to the success of Scott's now-famous "Anaconda Plan" would be the impatience of a northern public oblivious to the South's many dangers. "They will urge instant and vigorous action, regardless, I fear, of consequences," wrote the septuagenarian general. "That is, unwilling to wait for the slow instruction of (say) twelve or fifteen camps, for the rise of rivers, and the return of frosts to kill the virus of malignant fevers below Memphis."[23]

A veteran of two major wars, Scott knew the risks involved with sending unacclimated federal troops into a tropical environment during summertime. His concerns were echoed by another of Lincoln's military advisors, Quartermaster General Montgomery C. Meigs. During a war cabinet meeting held in June 1861, Meigs, a native Georgian, was asked whether the military should first conquer the Mississippi River Valley or launch an attack against P.G.T. Beauregard's army, which was then gathered near Manassas, Virginia. He told the group that he preferred to confront the rebels in northern Virginia rather than "go into an unhealthy country to fight them."[24]

The northern public was more concerned about the South's dangerous disease environment than Scott knew. Yankee physicians fretted about the consequences of sending U.S. soldiers "familiar with Northern climates and Northern diseases" into an area infested with dangerous "marsh miasm." In *Hints on the Preservation of Health in Armies,* first published in 1861, John Ordronaux warned that "Northern troops, in passing now no farther south even than the lower Chesapeake," would "enter a climate entirely foreign to their constitutions." He thought that since Union troops from "Northern seaboard States" would be most affected by the South's pernicious diseases, soldiers from "the great valleys of the Ohio and Mississippi" should garrison forts located in "lagoons," "low, alluvial regions bordering on the sea," and "the deltas of great rivers." The pro-Union press seemed more worried about the threat posed by yellow fever. One *New York Times* editor advised the War Department to seize and hold inland southern cities such as Milledgeville, Georgia, and Montgomery, Alabama, arguing that these locales were less susceptible to the horrific "yellow jack" outbreaks that plagued port cities such as Savannah and Mobile. The British media hyped the threat as well by speculating that the "climate" and "terrible yellow fever" of the South might allow the Confederacy to "defy . . . all the levies that the North can bring against it."

In nearly every area of the South Union troops would be forced to invade and occupy—the Gulf Coast, the tidewater region of the Southeast, the marshy lowlands along the Mississippi River—anopheline and *Aedes aegypti* mosquitoes thrived. Southerners were confident their "sickly season" would disrupt the operations of their unseasoned enemies but soon discovered that the diseases that they claimed as their own could work against southern armies as well. Richmond's need for soldiers brought young men together from the Mississippi Delta and the Virginia hills who had never been exposed to each other's plasmodium parasites, and during the first year of the war thousands of them fell ill with malaria. Yellow fever was also a threat to the Confederate army, a majority of whose members had grown up in rural areas and had never been exposed to the disease. As men from all over the country assembled to settle the issues of federalism and slavery on the battlefield, the mosquitoes of the South were galvanized by the large number of new prey that suddenly appeared in their midst. And before the guns fell silent, these tiny insects played a significant, and heretofore underappreciated, role in the events of the Civil War.[25]

The Glory of Gangrene and "Gallinippers"

For the average soldier who participated in the American Civil War, disease posed a greater threat to life and limb than enemy battalions. A staggering two-thirds of the roughly 360,000 northern soldiers who perished during the conflict died of disease, while Confederates suffered in equal, and perhaps even greater, numbers. Unfortunately, exact figures for the southern soldiers will never be known because most of their medical records went up in smoke along with their short-lived republic's dreams of independence and perpetual slavery. But after the war Joseph Jones, one of the Confederacy's most respected surgeons, argued that three-fourths of southern military deaths were attributable to pathologic causes.

Whatever the actual number was, it is clear from the surviving records that a battery of diseases stalked soldiers in both armies for the duration of the war. After the fall of Fort Sumter men who had enjoyed relatively good health on isolated farms in rural areas suddenly found themselves herded into unsanitary army camps that were teeming with bacteria, parasites, and viruses of every description. Mumps, measles, typhoid fever, gangrene, rheumatism, pneumonia, and chronic diarrhea were just a few of the maladies that beset both Union and Confederate troops and made military service a less than glorious enterprise. For Billy Yank and Johnny Reb the war was as much a story of putrid infections and burning fevers as one of long marches and frontal assaults.[1]

Malaria afflicted the common soldier more frequently than most other diseases. Among Union troops only diarrhea and dysentery were bigger problems. Statistics compiled by the federal government show that on average 224 out of every 1,000 northern men who sought treatment for an illness during the war were diagnosed with malaria. Confederate forces were also hit hard by the disease, which in 1861 and 1862 accounted for one-seventh of the cases of sickness among southern soldiers stationed in the East. Missing from these figures are the unknown numbers of men who either suffered without treatment or were diagnosed with another disease that completely overshadowed the symptoms of their plasmodium infection.[2]

Soldiers contracted malaria almost as soon as they joined the army. The disease plagued the numerous training camps that were set up early in the war by both the Union and Confederate governments. In the Washington, D.C., area mosquitoes buzzed in and out of the thousands of tents that were pitched within sight of the unfinished Capitol dome, spreading sickness among troops who had only recently arrived from the farms and factories of the North to fight for their country. Camp Michigan was located just a few miles from Alexandria, Virginia, and served as the headquarters for the Third Michigan Infantry and three other regiments. During a three-month period in 1861 George Willson, an assistant surgeon with the Third, had his hands full with patients but managed to record details of the thirty-six hundred cases of sickness that appeared in camp. Willson's notes show that "catarrh," a nonspecific Victorian term used to describe almost any ailment that caused sinus congestion, was the chief complaint among the troops, followed by diarrhea and malaria. More than five hundred men were diagnosed with either "intermittent quotidian fever" or "intermittent tertian fever." Bronchitis, rheumatism, and typhoid fever also tormented the soldiers.

In nearby Arlington the Twenty-fifth New York Volunteers, a predominantly Irish-American regiment recruited in New York City, occupied an unhealthy piece of ground they called Camp Smith. Bored and poorly disciplined, the men of the Twenty-fifth drank to pass the time and urinated within their campsite, even though a latrine had been dug only twenty-five feet away. At night groups of six soldiers huddled into one of the seven-by-five-foot "wedge tents" issued by the government. Anopheles mosquitoes that had hatched in the stagnant pools surrounding the Potomac were attracted by these clusters of warm-blooded bodies, and soon the soldiers of the Twenty-fifth were suffering from "intermittent fever," in addition to diarrhea and rheumatism. Other units posted near the Potomac River also contracted malaria. When one-third of the soldiers in a number of regiments camped on the Arlington "flats" fell ill with "diarrhea, intermittent, and typhoid fevers," the medical director of the Army of the Potomac, Charles Tripler, recommended that they be moved to higher ground. But Brigadier General Irvin McDowell's chief surgeon doubted that such a move would make any difference in the health of his men and openly expressed his skepticism in a report addressed to Tripler. Some units headed for higher ground anyway. The Fifth Michigan Infantry set up camp on Arlington Heights—outside the flight range of the myriad mosquitoes that infested the river bottomlands—and avoided plasmodium infections altogether (typhoid fever was still a problem for the Fifth). But for the most part Tripler was besieged on all sides by obduracy and incompetence. Many regiments arrived in Washington either without medical officers or with volunteer surgeons who had received only limited medical training and were "accustomed to a village nostrum practice." They disregarded army supply regulations and freely dispensed whatever medicines they thought might work at a time when the medical director was grappling with the logistics of moving medical stores to needy regiments.

This disorganization was caused in part by the haphazard selection process used for regimental surgeons in 1861. Some surgeons were appointed by state governors for political reasons, while others were elected by the soldiers of the regiment. Eventually, the federal government created examining boards that weeded out the most egregiously incompetent physicians and established a structured chain of command for the medical corps, but no such regulations were in place at the beginning of the war.[3]

Even the best-trained physician practicing in the 1860s, however, could do little to help his sick patients. The medical profession had made little progress since ancient times, and some of the most highly respected surgeons of the day still believed that diseases were caused by an imbalance of the body's four humors ("blood, phlegm, black bile, and yellow bile"), an idea first introduced by Hippocrates around 400 B.C. Unfortunately, Joseph Lister's experiments with antiseptics, Robert Koch's postulates on germs, and Walter Reed's discoveries regarding yellow fever were still years away, which meant Civil War surgeons could only treat symptoms rather than address the root causes of disease.[4]

In many cases untrained civilian volunteers did a better job of preserving the health of soldiers than surgeons did. The United States Sanitary Commission (USSC) was formed in New York City in 1861 by women and men who were worried that American armies might face the same epidemiological disaster that befell British forces in the Crimea. Not knowing that bacteria, parasites, and viruses were responsible for disease, they were nevertheless observant enough to make a connection between cleanliness and good health. Sanitary Commission volunteers published health advice for the army, lobbied government officials to place a higher priority on hygiene, donated equipment, and dispatched inspectors to campsites and hospitals to make sure that these areas were kept clean. Although the USSC was only an advisory body and could not enforce its recommendations, its high public profile goaded the government into enacting reforms that undoubtedly saved lives.[5]

Sanitary Commission volunteers worked in Union camps out West, where the situation was even more chaotic and unhealthy for the Federals than it was in the East. In Cairo, Illinois (the spot Dickens had dubbed "a breeding-place of fever" years earlier), the armies that would eventually triumph on the battlefields of Tennessee and Mississippi were then just a disorganized mass of sickly recruits, braying mules, and nonplussed officers. Surgeon John H. Brinton was medical director for the district and faced many of the same problems Tripler was dealing with in Washington. Brinton grew annoyed at the incompetence of commanders who seemed unable to keep their troops healthy and was surprised at the enlisted men's tendency to succumb to even mild ailments. "The men from the country had often not passed through the ordinary diseases of child life," he later wrote in his memoirs, "and no sooner were they brought together in camps, than measles and other children's diseases showed themselves, and spread rapidly." He also noticed that the "malarial influences" of the nearby Ohio and Mississippi rivers were sickening soldiers, especially those who hailed from "higher regions" and healthier locales. Unbeknownst to Brinton, many of these men were being infected with plasmodium parasites for the first time in their lives by the clouds of mosquitoes that hovered around the campsite.

"Violent remittent" and intermittent fevers appeared among the troops along with cases of typhoid fever, causing widespread sickness and death. In September 1861 the Eighth Illinois Infantry was camped in Cairo, close to a levee on the Mississippi River. Remittent and intermittent fevers soon broke out among the soldiers, who were also battling diarrhea and typhoid fever. Scores of these men were either sent to the regimental hospital or confined to quarters. Many of their comrades assigned to other locations in the West suffered a similar fate. Union camps in Kentucky and Missouri were plagued with malaria and other diseases. The Sixth Iowa Infantry left home to cheering crowds and fife and drum corps tributes only to discover that the realities of army life did not jibe with the heroic picture portrayed by politicians. A measles epidemic sickened close to one hundred members of the regiment while it was stationed in Jefferson City, Missouri. At the same time, "a large number of serious cases of chills, fevers, and camp diarrhoea" appeared among the men, putting a tremendous strain on the Sixth's medical staff.[6]

The Confederacy's armies were even sicklier. Like their northern counterparts, southerners who had grown up in sparsely populated areas were susceptible to ailments that nineteenth-century physicians considered "infantile diseases," such as whooping cough and measles. Measles proved especially problematic among new recruits, and several regiments were temporarily disbanded in order to check its spread. Meningitis, respiratory tract infections, dysentery, and typhoid fever were just a few of the other diseases that sickened southern volunteers early in the war. Malaria was also pervasive. The ailment was endemic in most areas of the Confederacy, and many men arrived in training camp already carrying plasmodium parasites in their bodies. Because different strains of the disease plagued different regions of the South, even recruits who had developed a tolerance for the variety of ague that existed in their home counties were vulnerable to bouts of chills and fever.

The prevalence of malaria in the Confederate ranks may help explain why southern troops often appeared sicklier to observers than northern soldiers. Edward Porter Alexander, who would earn lasting fame as an artillery officer in the Army of Northern Virginia, traveled to the North during the early months of the war and noticed that the Yankees he saw "were all fine healthy looking men, with flesh on their bones & color in their cheeks, thoroughly well uniformed, equipped & armed." In contrast, the Rebels he encountered in Georgia and Tennessee "were less healthy looking" and "sallower in complexion." By 1860 malaria had all but disappeared from New England and was on the decline in the northwestern states due to long winters and northerners' efforts to drain and develop their land. In the South, on the other hand, mosquitoes could survive most of the year, and southerners' dependence on slave labor produced crude housing, poor sanitation, and undeveloped land—conditions that aided the spread of plasmodium parasites. War brought men and malaria together, and the clouds of mosquitoes that were a part of the southern landscape quickly spread sickness through the ranks of both the Union and Confederate armies.[7]

Malaria was not the only mosquito-borne illness that threatened the health of the average enlisted man during the Civil War. Most Union and Confederate servicemen had never been exposed to yellow fever and harbored a deep-seated fear of the disease that had killed so many southerners in the 1850s. A few months before South Carolina seceded, William Tecumseh Sherman admitted in a letter to a colleague at the Louisiana State Seminary of Learning and Military Academy (later Louisiana State University) that his fellow northerners possessed "an utter dread of the yellow-fever and other epidemics of the South." His assessment proved accurate for the throngs of unacclimated Union soldiers who were tasked with invading and occupying the unhealthiest areas of the South. While anxious to prove their mettle in combat, many of these men became unnerved at the prospect of facing a disease they knew could wipe out an entire army in a short period of time. In New Orleans, for example, Major General Benjamin Butler watched in frustration as a number of officers in his occupation force were "frightened home by fear of yellow fever." The disease also preyed on the mind of Alfred Castleman as he marched with the Army of the Potomac through Virginia in the spring of 1862. Castleman doubted that the war would end even if the North captured Richmond because the Rebels could easily "fall back on the Cotton States" and lure the Yankees against defenses "more formidable than [Southern] guns," which he enumerated as "rice swamps, hot weather, and yellow fever." Several months later, in December 1862, navy man William Holton was thankful that he and his fellow sailors had avoided the disease, which he thought was "more to be feared by unacclimated persons than the fiercest battles." It is clear from these and other accounts that most Union military personnel considered the scourge of the South every bit as dangerous as Confederate cannons.[8]

Although they initially hoped that "yellow jack" would aid their cause by sickening unacclimated Yankees, southerners soon realized that the disease threatened their own armies as well. Early in the war the virus killed gray-clad troops in North Carolina and Texas, forcing the withdrawal of a number of garrisons protecting the strategic ports of Wilmington and Sabine. As the conflict continued, other outbreaks surfaced in South Carolina and Texas which claimed the lives of even more Confederates. While it was true that southern soldiers who had grown up in towns that were regularly afflicted with yellow fever often contracted the disease in childhood and acquired lifelong immunity, the bulk of the South's armies were composed of young men from rural hamlets that had never experienced an epidemic. For them the thought of entering an area plagued with yellow fever was just as terrifying as it was for their northern enemies.[9]

Both malaria and yellow fever were transmitted by the incalculable number of mosquitoes that Civil War soldiers encountered nearly everywhere they went in the Confederacy. The insects, often called "gallinippers" by Rebel and Federal soldiers alike, were active most months of the year in the states of the lower South. "The mosquitoes are becoming very thick out here," wrote Union captain Jonathan Johnson in April 1863 from his post near New Orleans. "We have the worst time here in the evening when we try to write. It is write with one hand and fight with the other and they say the worst is yet to come." The worst came for Dr. Junius Bragg in July 1863 while he was stationed with the Confederate army in Delhi, Louisiana. Bragg thought the mosquitoes he encountered there were "the largest, hungriest, and boldest of their kind." At night he covered himself from head to toe with a blanket to escape the pests but was bitten the moment any part of his body became exposed, however briefly. "May you never know a mosquito!" he wrote in exasperation to his wife, Josephine. As a surgeon in the southern army, Bragg saw countless cases of malaria and contracted the disease himself while stationed in his native Arkansas. William Wiley, a private with the Seventy-seventh Illinois Infantry, thought the "moskeetoes" he saw in southern Alabama were "as big as an Illinois wasp" and were "worse than rebel bushwhackers." The hands and faces of the men in his unit were soon covered with bites, and Wiley developed a severe case of malaria.

During the siege of Vicksburg soldiers became so frustrated by the hordes of mosquitoes they found inside their tents that they exploded powder cartridges in order to smoke the pests out. One enlisted man joked in a letter to his parents that an especially large Delta gallinipper had grabbed him "by the throat" one night and made off with his "Boots, hat, & 5,000 doll[ars] In Green Backs." Dr. Thomas Barton was also in Mississippi during the summer of 1863 and spent time along the Yazoo River, an area he described as "abounding with malaria." Barton found the local mosquitoes to be "very troublesome" and was shocked that some of the insects were "as large as the common house fly." They swarmed his tent at night, forcing him to move to higher ground on more than one occasion. In South Carolina the insects were still a nuisance as late as November. A week before Thanksgiving, Orra Bailey wrote to his wife from Beaufort, explaining why he thought slavery might be a necessity in the South—he believed he needed a minimum of three hundred slaves to keep the mosquitoes away.[10]

Picket duty was often the time when the insects were the biggest annoyance. Soldiers posted in wooded thickets and swamps at night were unable to light fires that might drive away the pests because doing so could give away their positions to the enemy. As a result, infantrymen frequently came off duty covered with red, itchy welts. While on picket duty in Louisiana, Henry Warren Howe was amazed to see clouds of mosquitoes that "filled the air like rain drops." He was equally surprised that his servant was able to fall asleep on the ground while so many of the pests hovered overhead. In South Carolina the men of the Ninety-seventh Pennsylvania also found the night air "filled" with mosquitoes while performing picket duty on James Island. Their comrades in the Seventeenth Massachusetts Infantry were pulled off picket duty in eastern North Carolina after swarms of anopheles infected most of the regiment with malaria. The troops of the Seventeenth were confined to the city limits of New Bern in order to expedite their recovery.[11]

Unbeknownst to Civil War soldiers suffering from malaria, the annoying but seemingly innocuous mosquito bites that appeared on their flesh were responsible for the violent chills and fevers they were experiencing. These trademark symptoms of the disease were especially worrisome to new recruits. One Illinois volunteer sent to Cairo for training in 1861 claimed that he and his comrades were "more afraid" of the ague in the area "than the enemy." The disease also shocked supposedly acclimated southerners. A Confederate stationed in North Carolina in the autumn of 1862 insisted that he was nearly shaken out of his garrison while suffering with a particularly bad bout of the disease. Veterans became accustomed to these "shakes" over time but still sought treatment whenever they reappeared, which was often. Troop transfers and the constant influx of fresh recruits meant new types of malarial parasites continually circulated through the ranks of both armies during the war.

Although *Plasmodium vivax* alone rarely proved fatal to soldiers, it created medical complications and weakened the immune systems of men who were already battling other, more dangerous diseases. Chester Rasbeck was a twenty-year-old private in the Tenth New York Heavy Artillery when he was admitted to Harewood Hospital in Washington, D.C., for "quotidian intermittent fever" and "chronic diarrhœa." Rasbeck's first week in the hospital was as hellish as any battle he had endured. He shook violently with chills and passed "fifteen to twenty stools daily." Doctors were eventually able to bring his malaria symptoms under control, but dehydration and exhaustion had already taken a lethal toll on Rasbeck's body—he died on October 26, 1864. John Fulton, a twenty-three-year-old private attached to the Sixteenth Pennsylvania Cavalry, also went to the hospital in 1864 for intermittent fever and chronic diarrhea. He was treated with a variety of medicines, including opium, but perished two days after Christmas. Even mild cases of malaria drained the strength of armies because soldiers sent to the sick ward for intermittent fever frequently never returned to their units. Many died after contracting more dangerous diseases in the unsanitary conditions that existed in the average Civil War hospital; others deserted or received unnecessary furloughs. During the early months of the war especially, inexperienced physicians granted medical discharges to "vast numbers of men" who were suffering from "diseases of a temporary character," a category that included malaria.

Oftentimes soldiers did their best to ignore their malaria infections in order to stay with their comrades. Civil War battlefields from Pennsylvania to Texas were traversed by men who were seriously ill with ague and any number of other maladies. During a battle in Virginia, John Kies's right forearm was punctured by a Rebel minié ball that fractured both his ulna and radius and created a wound that bled profusely. While receiving treatment for this injury, Kies confessed to doctors that he had been suffering from diarrhea and malaria for almost two months. His reluctance to seek treatment for these diseases could very well have stemmed from his fear of hospitals, a common phobia among soldiers during a time when a visit to a sick ward often turned into a trip to the morgue. In Kies's case his arm was successfully amputated, but he died several weeks later in his hospital bed after suffering with diarrhea and "a hard dry cough."[12]

Soldiers who did seek treatment for their malaria infections usually received quinine, a medicine derived from the bark of the South American cinchona tree. Civil War physicians were well aware of the value of cinchona, which by 1860 had been in use as a malarial specific for over two centuries. They prescribed quinine for a range of ailments, even though it was only effective against plasmodium parasites. Unaware that the drug suppresses the symptoms of malaria rather than prevents infection, surgeons on both sides employed it as a prophylactic. These experiments were successful when the amount doled out was large enough to check outbreaks of chills and fever. Many of Grant's troops, for example, benefited from regular doses of quinine during the sieges of Vicksburg and Petersburg. Soldiers in the armies of the Tennessee and the Potomac lined up to receive their daily dram of the drug, which was frequently mixed with whiskey to make it more appealing. Previous Union campaigns in these areas had failed in part because of a scarcity of the medicine. Despite recommendations from U.S. Sanitary Commission officials and others to dispense it regularly as a preventative, the Union medical department was too disorganized during the early years of the war to provide its surgeons with sufficient supplies of the drug. Compounding the problem were the quack remedies soldiers found advertised in northern journals and sold in camp by dishonest peddlers. Men who believed that "Dr. D. Evans's Medicated Flannel Abdominal Supporter" or "Hawes Vegetable Tonic" would cure their ague had no reason to subject themselves each day to the bitter taste of quinine. This chaotic situation—a reflection of the independent streak that was common among Civil War soldiers—allowed the disease to spread through the Union ranks unabated, sickening infantrymen who were needed on the battlefield.

The Confederate medical department suffered from quinine shortages for most of the war. The effectiveness of the Union naval blockade meant southern surgeons were forced to use crude, ineffective substitutes such as turpentine plasters and various indigenous barks to treat malarial fevers. Given the prevalence of malaria in the South, it is astonishing that any Confederate troops were healthy enough to fight by the end of the war, when Richmond's quinine supplies were extremely low.[13]

The only other effective weapon soldiers had to use in their fight against malaria was the mosquito net. These devices were issued solely to keep pests away but had the unintended effect of helping troops stay healthy. One purchase authorized in May 1863 by the U.S. surgeon general was for "two hundred mosquito netting frames, at $2.00 each," to be used in camps and hospitals. Troops were delighted to hang these "mosquito bars" in their sleeping quarters, which offered a modicum of relief from the South's superabundance of insects. One Union soldier stationed in Louisiana was optimistic that his new bar would "keep off the varmints completely." He and his comrades were given the nets for free but were told they would have to pay "$3.20 apiece" for any that were lost or ruined. Mosquito netting made life easier for Dr. Charles Brackett while he was camped along the Mississippi River near Helena, Arkansas, in August 1862. "There is not much comfort here after night expect under my mosquito bar," he wrote to his wife. "Once there I am safe. It is so nicely fitted, round my cot, hanging from the top of my tent." Brackett considered the item a novelty and suggested that he might "bring one home" to his native Indiana. Southerners who had grown up in mosquito-infested areas did not need a lesson on the importance of bars. When the Confederate government failed to issue one of the devices to a Texas soldier, his mother offered to send him her own. If she ever did, it was not soon enough to prevent an anopheles mosquito from injecting plasmodium parasites into her son's bloodstream.[14]

Despite the widespread use of mosquito bars and quinine, malaria plagued both armies during all four years of the war. It also sickened the thousands of Union and Confederate sailors who patrolled the South's inland waterways and coastal harbors during the war in steam-powered gunboats. These vessels, which contained dense concentrations of sweaty bodies, served as convenient feeding stations for the swarms of mosquitoes that hatched in pools along the shorelines. William Reiser experienced "frequent attacks of severe intermittent fever" throughout the summer of 1862 while serving on board the USS *Louisville*. His recurring chills and fever periodically kept him from his duties until he was eventually sent to the Union hospital at Mound City, Illinois, for treatment. John Winkleman, a nineteen-year-old German immigrant serving aboard the *Cincinnati*, also went to Mound City due to a plasmodium infection. He was successfully treated with quinine but developed a cough, severe back pain, and diarrhea while in the hospital, which resulted in his discharge from the navy on December 5, 1862. Erie, Pennsylvania, native Charles Johnson was a "coal passer" for the USS *Neosho* when he first began showing symptoms of malaria. His fever paroxysms left him so weak that the *Neosho*'s surgeon had no choice but to deem him "unfit for service." Confederate sailors were also plagued by malaria. The disease sickened southern seamen by the boatload and even disabled the James River Squadron at a time when it was involved in a desperate fight to keep Union forces from reaching Richmond.[15]

But no matter how much suffering malaria caused for Civil War sailors and soldiers, it paled in comparison to the agony endured by those who were unlucky enough to contract yellow fever. The disease flared along the coasts of North Carolina, South Carolina, Florida, and Texas during the war and burned through a number of U.S. Navy vessels patrolling the waters of the South Atlantic, the Gulf of Mexico, and the Mississippi River. Because the early symptoms of yellow fever mirrored those of a number of other maladies that were common among enlisted men, physicians, especially northern ones, frequently failed to identify the disease until it reached an advanced stage. A group of Union soldiers stationed in South Carolina, for example, was diagnosed with "a nonspecific bilious fever" but were in fact suffering from yellow fever. Doctors realized their mistake only after several of the men died. Patients infected with yellow fever generally experienced headaches, body aches, and fatigue before the telltale signs of the virus—black vomit and bleeding from the nose and mouth—appeared.

Charles Winters, an eighteen-year-old U.S. marine stationed near Mobile, Alabama, was diagnosed with remittent fever after he complained of a headache, fever, and severe pain in his back and abdominal area. Winters was also shaking with chills, and his tongue appeared "coated and dry" to the physician on duty. Within a matter of hours the young soldier was delirious with fever and vomiting blood uncontrollably. He never recovered. Winters's mate, a twenty-three-year-old Baltimore resident named Henry Gilman, was also misdiagnosed with remittent fever after he was "seized with a chill" and came down with a fever. Gilman was given massive doses of quinine and seemed to be improving for a short while before his condition took a turn for the worse. During the final few hours of his life he groaned in agony between bouts of vomiting and watched helplessly as his muscles twitched involuntarily. Gilman's surgeon was unable to do anything but "make him [as] comfortable as possible." He died on the morning of September 2, 1863, four days after his first symptoms appeared. Dr. D. W. Hand lived through an epidemic in North Carolina which killed hundreds of Union soldiers. He noticed that many of his patients first experienced chills, fever, restlessness, and severe back pain before becoming dangerously ill. Most of them also suffered from headaches, which were often concentrated in the area around their eyes. Dr. G.R.B. Horner's yellow fever patients at Key West, Florida, complained of headaches and pain in their loins. Several had glazed eyes and coated tongues and passed urine that Horner described as dark enough to stain wood.[16]

Nineteenth-century physicians responded to these symptoms with a variety of treatments, most of which did more harm than good. Hand and his medical team dosed yellow fever patients with calomel (mercury) until their gums turned blue. The doctors were unaware that the element they were using as a medicine is toxic to the human body and destroys healthy mouth tissue if ingested in sufficient amounts. Soldiers overdosed with calomel lost teeth, suffered hideous facial deformities, and sometimes died. Hand also experimented with "acetate of lead" and dispensed opium, morphine, and "iced sherry." To his credit he recognized that "careful nursing" was a better cure than any medicine, especially when the patient reached the final stage of the illness. Assistant Surgeon W. F. Cornick submerged his patients in hot baths "containing from four to eight ounces of mustard" until they were sweating profusely (or about to faint) and then put them to bed wrapped in blankets. The sick men were subsequently given calomel, castor oil, quinine, and "spirit of nitre" over the next several hours. During a major epidemic Dr. Nathan Mayer dispensed "very large doses of quinine" until he realized that the drug was not helping his patients. In desperation he turned to a remedy once used by English physicians in the West Indies, which called for alternating doses of calomel and castor oil. Such impromptu treatments were no more effective than the ones recommended by the government.

The Surgeon General's Office endorsed the use of "gentle diaphoretics" (agents that increase a patient's perspiration), lead, and "blisters or sinapisms" applied to the abdomen, in addition to quinine and mercury. The office also encouraged its physicians to borrow treatment methods from the personal accounts of surgeons who had lived through yellow fever epidemics. Washington never developed a central bureaucratic response to yellow fever because the best scientific minds of the day disagreed on the etiology of the disease. The U.S. Sanitary Commission had its own recommendations. A list of remedies published by the organization in 1862 included lead, creosote, and carbolic acid.[17]

Given the widespread use of such crude treatments, it is miraculous that any soldiers infected with yellow fever managed to survive. Their comrades who were suffering from malaria, on the other hand, had the advantage of quinine, one of the few drugs employed by Civil War surgeons which actually benefited patients. As a general rule, federal troops enjoyed greater access to the drug than their Rebel enemies, especially during the final months of the war, when the South was suffering from dire shortages of medicines as a result of the Union blockade. But even quinine was not enough to stop plasmodium parasites from sweeping through the ranks of both armies and weakening men who were already suffering from a battery of other diseases. In addition, malaria by itself could kill a soldier. "Congestive intermittent fever," most likely *Plasmodium falciparum*, was diagnosed in 1.2 percent of malaria cases among white Union troops and 2.1 percent of cases among black troops.

From the moment they showed up in training camp until they were discharged or died, Civil War soldiers were stalked by harmful microorganisms that thrived in nineteenth-century America's unsanitary conditions. Few diseases were more pervasive during the conflict than malaria, which was spread by the armies of anopheles mosquitoes which occupied nearly every region of the South. Whether stationed along the Atlantic coast, in the Mississippi River Valley, out West, or in one of the Gulf States, troops sweated and shook with fevers and chills caused by bites from these insects. These regions were also infested with *Aedes aegypti* mosquitoes, which had caused multiple yellow fever outbreaks during the antebellum period. Soldiers were annoyed by the clouds of gallinippers that swarmed their tents and tortured them while they were on picket duty, but they had no way of knowing that these pests would play an important role in several of the war's military campaigns.[18]

{3}

Mosquito Coasts

T HE MOB THAT gathered at the New Orleans waterfront on the morning of April 25, 1862, was not about to accept defeat graciously. It taunted and threatened the Yankee sailors who were hastily mooring their gunboats to the city's docks after blasting their way past a Confederate flotilla and two heavily armed forts. Behind the mob fires meant to keep cotton and other valuable commodities from falling into Union hands sent plumes of smoke billowing into the warm spring sky. But the demonstration was too little too late. Flag officer David Farragut's decision to run his fleet past New Orleans's defenses had proven to be a stroke of genius. The South's most important and heavily populated commercial port was once again under the control of the United States government. The North's plan to squeeze the Confederacy into submission with a snakelike naval blockade was working.

Even so, the rebellious spirit of the city was alive and well six days later, when an army of fifteen thousand bluecoats under the command of Major General Benjamin F. Butler arrived to take control. Butler's troops encountered the same menacing stares and angry insults that Farragut's men had endured. But while the residents of New Orleans were enraged over the Yankee occupation, they also held out hope that nature, or perhaps nature's God, might intervene on their behalf.[1]

They had good reason to be hopeful. Yellow fever outbreaks had swept through New Orleans nearly every year since 1817, killing more than forty thousand people and establishing the city's reputation as one of the unhealthiest places to live in the United States. The decade just before the Civil War was especially sickly. Back-to-back epidemics in 1853, 1854, and 1855, followed by another one in 1858, produced a staggering eighteen thousand casualties and virtually wiped out the city's nonimmune immigrant population. New Orleans natives knew that when "yellow jack" appeared in their midst, newcomers were especially susceptible to its ravages, giving rise to one of the malady's more fitting sobriquets, "Stranger's Disease." The thousands of Union soldiers tasked with maintaining order in the city were the perfect target for the virus because none of them had ever lived through an epidemic. Yellow fever might succeed where the Confederate military had failed and rid the city of its occupiers.

The southerner responsible for the defense of the region, Major General Mansfield Lovell, certainly hoped for such an outcome. Having ignominiously abandoned the Crescent City when Farragut's ships steamed up the Mississippi River, Lovell's contingency plan included the use of five thousand men to confine "the enemy to New Orleans, and thus subject him to the diseases incident to that city in summer." Southern newspaper editors gleefully predicted an outbreak that would soon destroy the unacclimated Yankee hordes and liberate lower Louisiana.[2]

Butler was certainly worried about yellow fever. He believed the locals were intentionally creating unsanitary conditions in the city in order to increase the likelihood of an epidemic, a scenario that made his men nervous as well. Residents preyed upon these fears. The city's schoolchildren chanted blood-curdling nursery rhymes in the streets about how the pestilence would soon wreak havoc on the Yankees, while parishioners with southern sympathies were said to be praying in the pews for the disease to arrive. Two locals even went so far as to measure and record the height of a group of soldiers; when asked to explain their behavior, they claimed to be taking measurements for the coffins that would become necessary once yellow fever burned through the Union ranks. These theatrics had some effect. A number of Butler's subordinates requested transfers or leaves of absence. Captain C. R. Merrill of the Fifteenth Maine Volunteers, for example, resigned his command when he suddenly realized after nine months of service that he did not possess "that peculiar military ability to command men, without which no one can be of much use in my present position."

Butler believed that the real motivation behind Merrill's resignation was his fear of yellow fever and asked the War Department to strike the captain's name from its record books.[3]

But punishing weak-willed officers did not solve the problem of how to keep the city healthy. After consulting with his medical staff and a few local doctors (some of whom were openly hostile), Butler decided that yellow fever was an imported malady that required local unsanitary conditions to survive. As a result, he chose to implement simultaneously the two strategies best known at the time for preventing the spread of the disease—a strict quarantine and fastidious sanitation measures. A quarantine station was set up seventy miles below the city and its officers given firm orders to detain any potentially infected vessels for forty days. In addition, a local physician was appointed to inspect incoming ships at the station and was threatened with execution if any vessels known to be carrying yellow fever were allowed to proceed upriver. These new rules caused a minor diplomatic row with the Spanish, who believed that their ships arriving from Cuba (where yellow fever was endemic) were being unfairly targeted for lengthy detentions. Butler assured Señor Juan Callejon, Her Catholic Majesty's consul in New Orleans, that he was not imposing "any different quarantine upon Spanish vessels sailing from Havana." To the relief of the State Department, Spain eventually dropped the matter but not before firing off a few strongly worded communiqués.

In town Butler put an army of laborers to work round the clock flushing gutters, sweeping debris, and inspecting sites thought to be unclean such as stables, "butcheries," and New Orleans's many "haunts of vice and debauchery." Steam-powered pumps siphoned stagnant water from basins and canals into nearby bayous. The northern press picked up the story and ran articles praising Butler's methods. "He will probably demonstrate before the year is out that yellow fever, which has been the scourge of New Orleans, has been merely the fruit of native dirt, and that a little Northern cleanliness is an effectual guarantee against it," predicted the editors at *Harper's Weekly*. The magazine published a cartoon five months later which featured the general holding a soap bucket and scrub brushes in front of an approving Abraham Lincoln.[4]

Butler's two-pronged approach ultimately proved successful; only two yellow fever deaths occurred within the city of New Orleans during the 1862 sickly season. His stringent sanitation policy eliminated many of the places where *Aedes aegypti* mosquitoes preferred to breed, such as rain-filled receptacles and stagnant pools, and the quarantine he enacted isolated infected insects and men. The Massachusetts general believed these actions preserved the health of the city, but residents were convinced that their enemies had simply gotten lucky. They held out hope that the next season would bring yellow jack and end their subjugation.[5]

That yellow fever might save the South was not a far-fetched idea in 1862, nor was it confined to the city of New Orleans. Butler's concerns over the disease were shared by the other Union commanders who were busy establishing beachheads and patrolling the waters along the Confederate coastline in response to Lincoln's April 1861 order to blockade southern ports. Outbreaks surfaced in Florida, North Carolina, South Carolina, and Texas during the first full year of the war which took a physiological and psychological toll on northern military personnel and disrupted Union operations. As events would show, the wall of mosquitoes which encircled the South's coastal areas sometimes proved to be a better defense barrier against Yankee incursions than Confederate cannon.

Such was the case in Texas. After seizing New Orleans, Farragut realized that his fleet could not effectively stop the illicit flow of weapons and supplies coming into the Confederacy through the Lone Star State without controlling the cities that offered smugglers safe harbor. When he learned in the fall of 1862 that Confederates were trading "hundreds of bales" of cotton through Matamoros, Mexico, Farragut wanted to seize Fort Brown and establish a Union troop presence in the area as quickly as possible. But his concerns about what yellow fever might do to an occupation force (the disease was "very bad on both sides" of the Rio Grande River at the time) and Butler's unwillingness to provide troops for the expedition kept Brownsville in Confederate hands for another year. Farragut seemed less concerned, however, about the health of his jack-tars. One of them, Henry French, was so worried by the cases of yellow fever which showed up on board the USS *Albatross* while it was off the coast of Texas that he ordered the ship back to the navy yard at Pensacola, Florida. Brownsville's Confederates were not surprised by French's abrupt departure. Soon after the Union commander's ship was spotted near Brazos Santiago Pass, one editor at the *Brownsville Flag* dared the enemy to enter his town. "Come on, Mr. Yankee," he taunted, "Fort Brown is garrisoned at present by Yellow Jack, and if you have a mind to try your strength with him, there will be no particular objections."

Unwilling to watch his entire crew fall prey to such a dangerous disease, "Mr. Yankee" set course for a healthier climate. Farragut was incensed at French's unauthorized retreat. "You were sent to blockade the Rio Grande, and you had no right to leave that station until regularly relieved by some other vessel," railed the admiral, who made it clear that for the navy at least, the importance of the mission trumped all other concerns: "Our duties in war are imperative, and we are as much bound to face the fever as the enemy, and to face both when necessary in our duty." An unlucky ship, the *Albatross* was again plagued with yellow fever the following year.[6]

French was not the only one of Farragut's officers in Texas waters who was worried about yellow fever. Brooklyn native William Renshaw received instructions in September 1862 to move his mortar flotilla "down the coast of Texas" in order to intercept blockade-runners and if possible, capture Galveston. Renshaw, inspired by Farragut's audacious seizure of New Orleans, ordered his fleet into Galveston Bay and demanded "the unconditional surrender of the city" after a brief exchange of fire with Confederate shore batteries. Renshaw's tough terms were softened, however, when he learned from two clever southern negotiators that yellow fever was raging in Galveston (in reality there was no epidemic in the city at the time). Not wishing to expose unacclimated Union troops to the disease, he agreed to a four-hour truce, during which time civilians could evacuate the city as long as Confederates did not bolster their defenses. Confederate colonel Joseph J. Cook took advantage of a loophole in these terms— there was never an explicit prohibition of the withdrawal of forces—to save several guns from capture by moving them to the mainland.

Renshaw was outraged at this apparent breach of faith and considered launching an attack in retaliation. But after considering the diplomatic implications of potentially killing the foreign nationals in the city, the strength of Confederate forces, and "the great danger of contagion from yellow fever," he allowed Cook to proceed without incident. Renshaw believed that "two old-fashioned 24-pounders, one 80-pounder rifle, and another gun" were not worth the price of exposing his sailors to a "fatal disease" that promised to kill "many innocent people." When the city was surrendered a short time later, he was content to lead a small force to raise the Stars and Stripes over the customhouse for thirty minutes and then retreat to the safety of his ship. His troops stayed close to Galveston's wharves. The guns and men Cook saved—in large part thanks to Renshaw's fears of yellow fever—were used by Major General John B. Magruder the following January to recapture Galveston for the Confederacy.[7]

When Renshaw learned that Farragut was having trouble scrounging up an occupation army for Galveston, he took two ships down the Texas coast and steamed into Matagorda Bay, an area infested with *Aedes aegypti* mosquitoes carrying yellow fever. After three people died and several others fell ill with the disease in September in the city of Matagorda, municipal authorities downplayed the significance of the deaths and confidently predicted that the worst was over. As evidence, they pointed to the diagnoses of two local physicians who were not convinced yellow fever was present (one had no experience with the disease, and the other only heard about the symptoms from patients who were mostly recovered). By October 22 their incompetence was exposed when at least thirty more people died, including a mother and her two daughters. Across the bay the virus was attacking the citizens of Lavaca at the same time. On October 31, 1862, Renshaw demanded the surrender of the city, which was defended by a small Confederate force under the command of Major Daniel D. Shea. Shea refused to give up but asked for time to evacuate the noncombatants, including many people who were ill with yellow fever. Because of Lavaca's epidemic, Renshaw granted the Confederate commander an extra thirty minutes (for a total of an hour and a half) to empty the town before his ships opened fire. Sick men and women hastily grabbed belongings and made their way to the railroad station to escape disaster. At least one ill man died along the way. When the shooting started, Rebel soldiers only partially recovered from yellow fever helped man the defense batteries and return fire. Shea's forces had been "decimated" by the disease but fought tenaciously to retain possession of the port town. In the town of Lavaca sick civilians unable to leave in the time allotted faced the twin scourges of war and pestilence as Union shells pocked the streets and damaged homes and businesses.

Although no one was killed during the bombardment, the Texas press later lambasted Renshaw for firing into a town of sick civilians. "The giving of an hour and a half to a community prostrated with yellow fever to evacuate the town, as was done at Lavaca, is of a piece with the savage enormities of our inhuman foe," railed one Houston editor. "It matters not to him if people are battling with a pestilence more deadly than his shells." He went on to predict that a "just God" would punish such "wickedness."

After a two-day bombardment Renshaw abandoned the expedition and left the city under Confederate control. Southerners were able to claim victory, and Shea was praised in the highest possible terms for his leadership. In reality yellow jack played a bigger role than the Confederates knew. Renshaw, who had been so reluctant to enter Galveston on account of yellow fever, was not about to expose his men to the same disease in a smaller, and strategically less important, town.[8]

But Texas's *Aedes aegypti* mosquitoes were without political sympathies and worked against southern forces in 1862 as well. In Sabine City, three hundred miles east of Lavaca, the Eleventh Battalion of Texas Volunteers (better known as "Spaight's Battalion") had a battle on its hands even before Union gunboats showed up in autumn. In July a British blockade-runner had arrived at the city's docks from the Gulf and disgorged along with its smuggled cargo either a sailor or insect carrying yellow fever. By September a full-blown epidemic was under way, with between three and six new cases of the disease appearing "every 24 hours." Before it ended, 150 civilians and a third as many southern soldiers were dead.

Panicked residents and military personnel alike evacuated the town. Those who fled to Houston and Beaumont were quarantined on the outskirts of both cities as Confederate authorities fretted about the possibility that the outbreak might spread. Mrs. Otis McGaffey lived in Sabine at the time and in later years recalled how her family's suffering was compounded by the egregious incompetence of local doctors: "[They] seemed to know little if anything about treating the patients, some getting drunk, and useless (as they were afraid of it)[,] others not knowing what to do." The McGaffeys fled to nearby Weiss' Bluff but were turned away by neighbors who were afraid of contracting the disease. Confederate officials dispatched Dr. George Holland, Dr. A. J. Hay, and several nurses to assist sick and dying patients, a number of whom were holed up in a local hotel that had been converted into a hospital. After a careful inspection of the town, Holland concluded that piles of "decaying fish and oysters" had contributed to the epidemic, which he believed stemmed from a combination of local and imported causes. He recommended a ban on all traffic coming in and out of the city in order to prevent the further spread of the illness, a decision that added food and supply shortages to Sabine's list of problems. Most of the Confederates occupying Fort Sabine were moved out of town to escape the plague. By the last week of September a Union squadron consisting of three vessels—the *Henry Janes,* the *Kensington,* and the *Rachel Seaman*—arrived and opened fire on the fort, then commanded by Confederate major Josephus Irvine. Realizing his position was untenable, Irvine ordered the guns spiked and withdrew his skeleton force by eight o'clock on the morning of September 25, 1862.[9]

The skipper in charge of the Union squadron that drove off Irvine's Confederates was forty-one-year-old Frederick W. Crocker, a former whaler from Massachusetts who had sailed around the world before joining the navy. Crocker soon realized his enemies had retreated and went ashore to raise the Union standard over Fort Sabine. One of his subordinates, Lewis Pennington, who commanded the *Henry Janes* and had once lived in Sabine, entered the city and learned from the few dozen residents left that yellow fever had killed "nearly one-half of the population" (including the town's mayor) and that the disease had forced several hundred Rebel troops to withdraw to nearby Beaumont. Pennington and Crocker were shocked to find the dreaded "scourge of the South" raging in Sabine and kept their men "close to their boats" as much as possible during the brief occupation, only sending them ashore to perform necessary tasks such as capturing a family of Confederate spies and burning the town's railroad depot. Thirty miles away, in Beaumont, the Confederates in Elmore's Twentieth Texas Regiment, who had been sent as reinforcements to drive the Yankees out of Sabine, were terrified at the prospect of entering an area plagued by yellow fever. The lieutenant colonel in charge of the regiment raised such a ruckus about the danger of the disease that he was ordered not to mention the phrase *yellow fever* in the presence of his superior officer. Equally uninterested in occupying a plague town, the Federals moved into Taylor's Bayou, ten miles away, and fired a railroad bridge (which was temporarily saved by an alert Confederate private who doused the flames, though Union troops later returned to destroy it). Crocker's forces continued to operate with impunity for another month, capturing prisoners and burning blockade-runners, before rumors of a massive Confederate attack unnerved Union commanders, who resumed their blockading duties and allowed the forts along the Texas coast to fall back under southern control.[10]

Crocker's bloodless conquest of Sabine was made possible by the swarms of *Aedes aegypti* mosquitoes that inhabited the city. One resident remembered that the pests were so numerous in the area that they "actually killed young calves and chickens" and tortured visitors, who were forced to cover themselves from head to toe in order to keep from being bitten. Yellow fever had killed several dozen Rebel soldiers and driven away those who remained to defend the area. In late September Union gunboats showed up and, together with a measles epidemic that was under way at the same time, helped keep the Confederates from mounting a counterattack. Shortly after the Federals attempted to burn the bridge at Taylor's Bayou, the Eleventh's namesake commander, Lieutenant Colonel Ashley W. Spaight (a lawyer and Alabama native who had moved to Texas only a year earlier), wrote to his commanding officer from Beaumont, describing the poor health of his troops: "Half of one of my infantry companies are down with measles and quite a number of two others [are] not yet entirely convalescent from yellow fever, which renders me short of men to guard all points and do the work on hand." A few days later he was asking for reinforcements to replace the large number of men who had become "unfit for duty" from both diseases. Spaight was never able to coordinate an effective response to the Union invasion, and Crocker's forces continued their campaign unmolested.[11]

Rumors of a yellow fever epidemic in Houston also emerged in the fall of 1862. When a handful of people in the city were diagnosed with the disease in October, a "stampede" of frightened residents fled to the countryside. They were shunned by neighboring communities seized with fear. In Richmond, Texas, authorities banned travelers from Houston from staying in the city for longer than a half-hour or from disembarking from railroad cars without an armed escort. Hempstead's city council prohibited evacuees from remaining in town longer than three hours and banned the importation of woolen goods from infected areas (the material was thought to easily absorb and transmit yellow fever). In reality the Houston outbreak never reached epidemic proportions. Although thirty to forty people contracted the disease, fewer than a dozen died from it. In late October freezing temperatures halted the area's mosquito activity.[12]

As these examples show, yellow fever raged all along the Texas coast during the autumn of 1862. From Sabine Pass to Brownsville *Aedes aegypti* mosquitoes helped create a military stalemate in the Lone Star State which ultimately worked to the South's advantage. Although the

U.S. Navy outgunned the Confederacy's shore defenses, the task of occupying the major trade hubs along six hundred miles of coastline proved impossible for the North during the sickly season, both because troops were needed elsewhere and because sending unacclimated northern military personnel into areas known to be plagued with yellow fever was an unwise and potentially dangerous strategy. Farragut, French, Renshaw, and Crocker were all unnerved at some point by the prospect of exposing their men to yellow jack and either withdrew their forces or altered their plans in the region.

A similar scenario unfolded in Florida during the same sickly season. The federal government had maintained control of the Keys after Florida seceded and by 1862 was using Key West as a coaling station for its blockading squadron. A yellow fever epidemic began there when the English bark *Adventure* sailed up from Cuba and landed in distress at Key West on June 20, 1862. Four crewmen showing symptoms of yellow fever were immediately sent to the Marine Hospital, where one of them died, while the ship stayed in quarantine for nearly two weeks. The virus lay dormant until the end of July, when a soldier with the Ninetieth New York Infantry suddenly succumbed to the disease. Soon workmen building nearby Fort Taylor began to die, and by August a full-blown epidemic was under way, which spread to the Union ships and merchant vessels anchored off the Florida coast. The first case the fleet's surgeon, Dr. G.R.B. Horner, saw was "a young black refugee" named William Talbot, who was on shore leave from the East Gulf Squadron's flagship, the USS *San Jacinto*, in late July. After spending some time in Key West, Talbot began suffering from diarrhea and a fever and was admitted to the Marine Hospital, where he was misdiagnosed with typhoid fever until he ejected black vomit. He died from yellow fever on July 30.

Another of the *San Jacinto*'s crewmen experienced vertigo, a partial loss of vision, and a "severe headache" while on furlough. Somehow he managed to get back to his ship but then developed a high fever, rapid heart rate, and nausea. Although he was fortunate enough to recover, the man's eyes remained "jaundiced for weeks." By August 1 the *San Jacinto* reported nine cases of yellow fever on board (two sailors had already died) and was ordered north to Boston to escape the plague. The ship's commander, Admiral J. L. Lardner, moved his flag to the *St. Lawrence*, but neither he nor the crew could escape the pestilence. By the middle of October Lardner was too sick to give orders. Other ships under his command—the *Dacotah*, the *Huntsville*, the *Rhode Island*, the *Tahoma*, and others—were also infected. Horner counted twenty-four vessels with yellow fever. At a time when northern merchants were writing letters to the secretary of the navy, Gideon Welles, asking for protection from Confederate privateers such as the CSS *Alabama*, the navy could ill afford to have so many vessels out of commission or operating with undersized crews.[13]

Onshore the situation grew dire. In August 153 soldiers stationed in Key West contracted yellow fever, including members of the Forty-seventh Pennsylvania Regiment. In September another 137 fell ill. An unknown number of civilians got sick as well. Surviving accounts suggest that the disease mostly afflicted unacclimated officers' families and northern workers who were newcomers to the city. One eyewitness claimed that Key West was one "vast hospital" for two months and that 200 deaths occurred within a matter of weeks, while a correspondent for the *New York Herald* reported 160 people dead a month before the epidemic ended. An outbreak of dengue fever, or "breakbone fever"— another virus transmitted by *Aedes aegypti* mosquitoes—preceded the yellow fever epidemic. Several soldiers and their families endured its ravages, suffering from intense headaches, rashes, and a painful crushing sensation in their bones.[14]

As cannons roared in Key West and flaming barrels of tar belched black smoke into the humid air, infantrymen in the Ninetieth New York stationed at nearby Fort Jefferson on the Dry Tortugas Islands began to drop like flies. Lewis Tucker was a twenty-six-year-old book-keeper from Walworth, New York, who fell ill at the end of August and died just a few days after the one-year anniversary of his enlist-ment. George P. Crocker was a blue-eyed, brown-haired "vender" who disliked army discipline. Initially given the rank of corporal when he enlisted in September 1861, he was demoted for "bad conduct" by the following January. Yellow fever killed him on September 20, 1862. His comrade, William Gage, was a tall, dark-haired clerk from New York City who had left his job and joined up with the Ninetieth in Brooklyn. He was only nineteen when a mosquito bite in August ended his life.[15]

As the epidemic raged, the navy began diverting ships away from Key West. Those that had to make brief stops to deliver medicines or pick up important messages were ordered to stay anchored offshore and smoke any "documents, letters, or articles" sent out in dinghies.

The whole ordeal created an enormous headache for Secretary of the Navy Welles, who already had his hands full stopping blockade-runners smuggling southern cotton to the West Indies and Rebel steamers that were torching northern merchant vessels. His headache ended when the fever burned itself out in November. But by that time seventy-one soldiers and an unknown number of civilians were already dead.[16]

Farther north, along the Florida panhandle, *Aedes aegypti* and anopheles mosquitoes were also tormenting soldiers and civilians. In June 1862 dengue fever appeared among the Union garrisons occupying Pensacola and was "a troublesome complaint" for "the remainder of the season." Physicians worried that the disease portended a yellow fever outbreak, which was rumored to be raging in nearby Mobile. In reality Mobile's residents would only narrowly avert an epidemic in September, when the CSS *Florida* arrived from Cuba at the end of a harrowing high seas chase involving a score of Union blockade ships. The crew was still recovering from a severe bout of yellow fever that had broken out while the vessel was in the West Indies. Even the ship's commander, John Maffitt, was so ill he "could hardly stand," having forced himself off of his sickbed in order to guide the *Florida* into port. Although infected ships running the blockade were responsible for the epidemics that were occurring in other areas of the South, the citizens of Mobile seemed oblivious to the danger they were in and showered Maffitt with flowers, "jellies, cakes and delicacies" for his heroic escape. Luckily, their city escaped an outbreak.[17]

Meanwhile, the epidemic in southern Florida spread to the Sea Islands of South Carolina, which had been captured in November 1861 by Union forces searching for a naval fueling station between Florida and Virginia. At the end of the summer the USS *Delaware* arrived at Hilton Head from Key West carrying a detachment of soldiers from the Seventh New Hampshire Infantry. They were "broken down in constitution," having spent too much time in the Florida heat. The entire Seventh had enjoyed the early part of the summer on the Dry Tortugas Islands "rowing for exercise," "fishing," and "cooking gull eggs" while ostensibly guarding a Union supply hub, but the men aboard the *Delaware* were suffering from an unidentified fever when they arrived in August. Because yellow fever was known to be raging in Key West, the sick soldiers were quarantined for twelve days before being sent to a nearby military hospital for treatment. The doctors on duty diagnosed "ordinary bilious fever of a mild type" until one private suddenly expired, and an autopsy revealed black vomit in his stomach. Other men soon began dying as well.[18]

A few days later, however, the department's medical director was hopeful that the disease had burned itself out. It had not spread to any other regiments, and the surviving men of the Seventh had been transferred out of the area. This information was passed on to the commander of the Department of the South, Major General Ormsby MacKnight Mitchel, who felt confident that he could proceed with his plans to harass enemy forces operating near Charleston and Savannah without further interruption. Mitchel, who had been a nationally renowned astronomer and math professor at Cincinnati College before the war, possessed a razor-sharp intellect as well as a reputation for taking risks. His audacious invasion of northern Alabama several months earlier and bold but unsuccessful attempt to sever the southern railroad linking Chattanooga and Atlanta (an event later dubbed the "Great Locomotive Chase") captured the attention of Union officials. From Washington's perspective he was just the sort of man who could get the ten thousand northern soldiers stationed on the Sea Islands into the fight. Mitchel urged the War Department not to let concerns over disease keep it from sending reinforcements. "The health of the officers and soldiers I find to be generally good," he wrote General-in-Chief Henry Halleck on September 20.[19]

After dark on October 9 Dr. Thomas Smiley arrived at the Hilton Head hospital and examined a patient from the Quartermaster's Department who was complaining of a headache, nausea, and abdominal pain. By candlelight Smiley could see that the man's face was "bloated and red" and that his eyes were "deeply congested." Learning that his patient was coming off a three-day drinking binge, the doctor assumed the problem was alcohol related and went home for the night. When he returned the next morning, he was surprised to find the man violently ill and spewing black vomit. Within a matter of hours his patient was dead. A few weeks later the virus that killed him spread through Hilton Head, causing intense anxiety, suffering, and death. In the end forty soldiers on the island would contract yellow fever, and twenty-five of them would die from it. Those who did not get sick were seized with fear. Men who had proven themselves fearless in combat worried incessantly about coming into contact with the sick. Union supply vessels stopped making deliveries, and mail service ceased. The supply disruption even caused suffering for the horses, which were let loose to graze due to a lack of forage. By one soldier's account the department "was virtually isolated."[20]

Mitchel's soldiers were not the only northerners in the area at risk. In the wake of the Union's military successes along the Atlantic coast, the U.S. Treasury Department had allowed northern benevolent organizations to dispatch teachers and humanitarian relief aid to blacks living on the Sea Islands of South Carolina. Most of the instructors attached to these groups were women who had never been exposed to either yellow fever or malaria but were willing to hazard their health in order to assist former slaves. When Charlotte Forten, an African-American schoolteacher from Philadelphia, arrived in Hilton Head in late October 1862, the yellow fever epidemic started by the Seventh New Hampshire Infantry was still raging. She and her party "immediately" took a boat to the healthier town of Beaufort (an antebellum summer resort for wealthy Charlestonians), where Union soldiers, feeling antipathy toward any program designed to educate blacks, warned them "that the yellow fever prevailed to an alarming extent" and that "the manufacture of coffins was the only business that was at all flourishing at present." Forten refused to be intimidated by these reports and remained in the area for over a year. In that time she experienced periodic bouts of sickness and was eventually forced to return north until the "unhealthiest season" had passed.[21]

Local blacks warned the philanthropists that the summer and fall months in the Sea Islands were dangerous for whites, but many chose to stay anyway. Arthur Sumner wondered why all the northern men who lived together in a house close to a swamp were sick with the local fevers. Sumner's confusion stemmed from his belief that the summer "sun and wind," both of which were blocked by a cluster of trees next to the house, were the chief threats to the men's health. Laura Towne, a teacher and nurse, dismissed warnings from slaves on St. Helena who told her whites needed to go "North or to Beaufort in summer." She thought her proximity to the ocean's breezes would keep her safe and believed the danger was exaggerated. A number of her black patients developed "chills and fever," which she considered "easily managed." Towne herself managed to stay healthy.[22]

Others were not so fortunate. One volunteer, who also thought the disease threat was overstated, became "dangerously ill" and was forced to return north for several weeks to recover. Samuel Phillips, the nephew of abolitionist Wendell Phillips, died of disease during his first sickly season in the South. Typhoid fever and other maladies worked in conjunction with the region's endemic malaria to wreak havoc on the health of the unacclimated newcomers. New England missionaries were living and working alongside locals who had been repeatedly infected with plasmodium parasites, which were easily spread by the swarms of mosquitoes that inhabited the Sea Islands in the 1860s. Missionaries were shocked by the numbers of these insects they encountered. "The mosquitoes are so horrible that we cannot, generally, write at all in the evening," complained Laura Towne in August 1862. She gave up trying to pen letters underneath the security of a mosquito net after twice setting it on fire with her candle. Another volunteer was miffed that the insects found a way underneath her netting and constantly woke her "with their singing and stinging." Luckily, the volunteers discovered that moving their quarters closer to the ocean improved their health. Stiff sea breezes and saltwater kept anopheles swarms at bay and provided the northern civilians with a respite from the plasmodium parasites that plagued the interior of the islands. Even so, malaria, together with other ailments, helped limit the effectiveness of the Port Royal experiment by either killing volunteers outright or forcing them to flee north to recover their health.[23]

Ormsby Mitchel never managed to escape the pestilential atmosphere of the Sea Islands. The general watched in anguish as four of the five members of his personal staff fell ill during the Hilton Head yellow fever outbreak and decided to move his headquarters to nearby Beaufort with the hope of avoiding more deaths. Mitchel himself soon showed symptoms of yellow fever, however, and after four days of tossing and turning in his bed in agony, stoically resigned himself to his fate. "The struggle of death has passed," he told an aide, "God has called me, and I cheerfully obey the summons." The general's passing brought mourning from the troops, who were convinced they had lost "an efficient leader" and "distinguished commandant." Flags flew at half-mast, gun salutes thundered at every post in the department, and the officers of the Tenth Corps wore mourning badges for thirty days. In the absence of an equally aggressive commander, the department lapsed back into apathy. Mercifully, the epidemic was over by the beginning of November.[24]

But as Mitchel lay dying, a much worse yellow fever epidemic was killing hundreds of people nearby in Confederate-controlled Wilming-

Kate steamed up the Cape Fear River loaded with "bacon and other food supplies" and disgorged its smuggled cargo at the city's waterfront. The *Kate* was arriving from Nassau, where a yellow fever outbreak was under way, and had at least one sick sailor on board. Other crew members soon fell ill, along with a few of Wilmington's residents. Rumors spread faster than the disease, which was said to be "raging" in the city in early September.[25]

Hoping to stave off a public panic, the city's mayor launched an investigation into the matter and was relieved to discover that only five people had died. He and the editors of the local newspaper believed the rumors of an epidemic were "exaggerated" and confidently informed their fellow citizens that the worst was over. By the end of the September, however, the number of cases mushroomed, and Wilmington's physicians were quickly overwhelmed with patients. Panicked families hastily gathered up belongings and fled to the countryside. Those who remained in the city were disproportionately poor or black. The De Rosset clan, whose patriarch was a successful Wilmington businessman, evacuated the city but left behind a number of their slaves. Several of these men and women contracted yellow fever, and one, a literate slave named William, wrote to his masters about the suffering they were enduring at home. "I hav never bin [so] sick in my Life Bee fore," he reported. Ill for over a month, William was forced to nurse other sick slaves while battling his own infection. He found solace in his faith and believed his life had been spared "for sum good purpis." The high rate of sickness among Wilmington's slaves must have surprised many whites who had previously believed that blacks were immune to yellow fever. Before the epidemic ended, as many as 150 African Americans had died.[26]

A number of physicians, ministers, and nurses risked their lives to provide comfort to the sick and occasionally fell prey to the disease themselves. An Episcopal priest named Robert Drane, one of only a handful of clergymen who remained in Wilmington during the epidemic, died of a mosquito bite after ministering to scores of hapless yellow jack victims. Drane's untimely demise prompted an outpouring of grief from Wilmington's residents, who sent his son letters expressing their gratitude for the rector's gentle counsel. Another casualty was James Dickson, a local doctor who also died in the line of duty. Dickson's peers considered him "a physician of the highest character and standing," and the community mourned his loss. Volunteers from South Carolina also put themselves in harm's way. Nuns from the Convent of Our Lady of Mercy in Charleston served as nurses, while P.G.T. Beauregard, who had only recently come back to South Carolina after serving in Mississippi, sent several of his surgeons to dispense advice and medicines.[27]

The streets of Wilmington were silent except for the crackling of countless fires that burned "in every street and yard" to ward off the pestilence and produced a thick cloud of smoke which hung over the city. The landscape was dotted with ponds created by heavy summer rains, which had also filled cellars and the spaces beneath floorboards with stagnant water. Train service stopped, hotels and businesses were shut down, and trash of every description littered the city's wiregrass lots. Behind closed doors sick men and women endured a level of suffering which was shocking even during an era when disease-related fatalities were common. Georgia Weeks was staying at "a house of ill-fame" when she first showed symptoms of yellow fever. She experienced headaches, back pain, and profuse vomiting for two days before a physician arrived. Unfortunately, the doctor could do nothing to alleviate her agony, and she died the morning after his departure. Robert Smith was a butcher who had been drinking for several days when he first felt sick. During the final twelve hours of his life Smith's skin turned yellow, his hands and feet became cold as ice, and he ejected black vomit. Another businessman named Dix decided to stay in town and face danger rather than allow his operation to shut down. His partner, Donald MacRae, fled the city but agreed to let the firm remain open. In early October MacRae sent a letter listing a number of items for Dix to purchase, which also contained a sober caveat. "I do not wish you to let these or any other orders induce you to expose yourself to the yellow fever," he warned. "I am still of [the] opinion that you had better keep out of town until frost." Unfortunately, MacRae's words were not enough to prevent an infected *Aedes aegypti* mosquito from sending Dix to an early grave.[28]

When Brigadier General William H. C. Whiting arrived in mid-November 1862 to take charge of Wilmington's Confederate forces, he found the city's defenses in disarray. Rebel troops had evacuated the town itself, leaving "many important points neglected." Whiting feared that the Federals would take advantage of the chaos to launch an attack the moment the first frost appeared. "Preparations have

been suspended, the garrison reduced and withdrawn, the workshops
deserted, transportation rendered irregular and uncertain, provisions,
forage, and supplies exhausted," he worriedly wrote to Secretary of
War George Randolph. "Unless therefore more speedy measures for re-
enforcement and relief be adopted I have great apprehensions of a successful *coup de main* on the part of the enemy." U.S. Navy commander
William Parker had just such a notion in mind and wanted to launch
an attack in December while the Confederates were still weak, but his
commanding officer, Major General John Foster, was more interested
in leading an expedition against Goldsboro, with the objective of destroying the Wilmington and Weldon Railroad, Richmond's link to the
Carolina coast. Of course, it is entirely possible that Foster was simply
unnerved by Wilmington's epidemic. Two years later, as commander
of the Department of the South, he refused to enter Charleston Harbor
under a flag of truce when rumors circulated that a yellow fever epidemic was raging in the city.[29] Had Parker's plan been implemented,
Wilmington might have fallen under Union control in the winter of
1862. Instead, it served as an important Confederate supply hub for
two more years. Whiting claimed that "even one or two" Union vessels would have posed a serious threat to the city in the immediate
aftermath of the epidemic.[30]

By the end of November cold weather halted the activity of the
city's *Aedes aegypti* population, and the epidemic ceased. Emotionally
exhausted survivors wondered if Wilmington would ever be the same.
"What a change there will be in our community," wrote one resident
who had fled to Greensboro to escape the pestilence. He was not sure
if the town could ever find "peace and quiet" again and lamented the
painful memories that would linger for years to come. The *Wilmington Journal*, which had temporarily suspended operations during the
outbreak, recognized that it would take a while before the town could
"recover from the effects of the destroying pestilence." Mary Ann Murphy's brother returned to Wilmington in November and was greeted
by both black and white residents who were glad to see a familiar face.
After exchanging pleasantries, they told him that more people had died
than the papers reported. According to the statistics compiled by the
Journal, 650 of the roughly 4,000 people who remained in Wilmington
during the epidemic died. Other sources place the number as low as
450 and as high as 1,200. Whatever the exact figure, the residents of
Wilmington were traumatized by the event. For many of their friends

and family members death came not from the muzzle of a Yankee rifle but from the proboscis of a female mosquito. Their plight also sent shockwaves through Washington which reverberated for the duration of the war. Both the Wilmington epidemic and one that simultaneously struck Hilton Head, South Carolina, sent Union military planners a strong message about the dangers of the Carolina sickly season.[31]

The Wilmington epidemic was the largest of a series of yellow fever outbreaks that occurred along the South's coastline during the sickly summer and fall of 1862. Union naval commanders tasked with implementing Winfield's Scott's "Anaconda Plan" found themselves battling a disease that was as terrifying as anything the Confederacy could throw at them. In Texas *Aedes aegypti* mosquitoes created a biological shield that helped thwart the navy's plans to seize southern ports, a task deemed necessary to shut down the illicit international trade that supplied Confederate armies. The same insects attacked Union garrisons in Florida and South Carolina and even killed the commander of the Department of the South, Major General Ormsby MacKnight Mitchel. Butler's sanitation and quarantine measures, in addition to luck, allowed the nonimmune northern troops stationed in New Orleans to avoid the pestilential nightmare that southerners had predicted would send them to an early grave. But occupying coastal port towns was not enough to bring the recalcitrant South to heel. Victory could only be achieved by sending U.S. troops into the southern interior to occupy Confederate territory and crush its armies. Millions of anopheles mosquitoes awaited the Federals.

{4}

"THE LAND OF FLOWERS, MAGNOLIAS, AND CHILLS"

WHILE FLAG OFFICER David Farragut was still getting his fleet into position below New Orleans, newly hatched female mosquitoes in western Tennessee were feasting on the Union soldiers following Major General Ulysses Grant up the Tennessee River and into the heart of the South. Grant's destination was Pittsburgh Landing, a riverside village located just north of the Tennessee-Mississippi border which Henry W. Halleck, the new commander of the Union armies in the West, wanted to use as a launching point for a thrust against the Confederates occupying the important railroad hub of Corinth, Mississippi. Halleck had ordered Don Carlos Buell's army, then occupying Nashville, to rendezvous with Grant's forces in order to create a massive northern juggernaut that would ensure a Union victory. But before the two armies could connect, Confederate general Albert Sidney Johnston launched a preemptive strike that produced a staggering twenty thousand casualties in two days of vicious fighting. Once the Battle of Shiloh was over and Confederates were beating a hasty retreat to the relative safety of Corinth, Mother Nature continued the killing.

Typhoid fever, malaria, diarrhea, dysentery, and scurvy broke out among the Union ranks, causing tremendous suffering and death. Dr. J. S. Newberry was a member of the U.S. Sanitary Commission in April 1862 and in later years recalled the pathetic scenes of human misery he had witnessed at Pittsburgh Landing. "The pestilential atmosphere of the country about Shiloh was producing an amount of sickness almost without parallel in the history of the war," he wrote. Newberry and his colleagues worked diligently to evacuate the sick and wounded on hospital ships and provide comfort to ailing northern soldiers. Diarrhea and dysentery were responsible for most of the suffering, but the presence of malaria caused additional medical complications. John Vance Lauderdale, a contract surgeon, knew firsthand about the malaria problem at Pittsburgh Landing. While serving on a Union hospital boat there, he began suffering from regular bouts of "ague." His fever and chills were present "every day, & a good deal every other day," making his time on the Tennessee River "drag very heavily."

As Halleck cautiously moved his army toward the Confederate forces at Corinth in May 1862, plasmodium parasites compounded the suffering of large numbers of his troops, who were dehydrated and disoriented by chronic diarrhea. Dr. T. H. Walker, an army physician, observed the pernicious effect the local fevers had on soldiers who were already sick with other diseases. According to Walker, diarrhea "prevailed as an epidemic throughout the camp," and "the majority of cases" he saw were "complicated with Intermittent Fever, producing great prostration and difficulty in treatment."[1]

Plasmodium vivax by itself was usually not fatal, but in conjunction with the various other maladies encountered by the average Civil War soldier, it could mean the difference between life and death. From April to June 1862 anopheles mosquitoes poisoned the blood of Federal troops who were already exhausted and sickly. In June the Army of the Tennessee, with an average strength of 66,042 soldiers, reported 2,541 cases of "intermittent" fever and 1,574 "remittents," with the numbers climbing each month through the end of October. These numbers, when added to the disease statistics from the Army of the Mississippi and a few other units (the numbers from the Army of the Ohio are not available), show that malaria was the second most frequent diagnosis in Halleck's command; only diarrhea and dysentery were more ubiquitous. Thus in May 1862 William T. Sherman, who was not yet a household name, had 10,542 troops on paper, but only 5,289 of them were fit for duty. More than 2,500 of the absent men were too sick to serve. Sherman himself supposedly contracted malaria while destroying Confederate supply trains near Chewalla, Tennessee, just north of Corinth. The soldiers of the Fourteenth Michigan Infantry were camped near Corinth and began experiencing "chills and fevers" following a heavy rainstorm that created a plethora of freshwater puddles in the vicinity. The soldiers of the Fourteenth had abandoned their "blankets and overcoats," along with their tents, during the march through Tennessee and were vulnerable to mosquito bites day and night.[2]

The Confederate troops holed up inside Corinth were even worse off, primarily because of the dirty drinking water and spoiled food they were forced to subsist on. Kate Cumming was a volunteer nurse during the siege who worked tirelessly to ease their suffering. She learned from a Confederate chief surgeon that almost all of the regiments in the city were ill and could easily be defeated if they were attacked. Cumming and "quite a number of the ladies and doctors" at Corinth soon got sick as well. When the town's wells ran dry, thirsty soldiers were reduced to slurping water from polluted puddles that doubled as cesspools. Dysentery and typhoid fever were rampant, and a lack of vitamin C added scurvy to the list of medical problems. In April 1862 Confederate general P.G.T. Beauregard had taken command of the western army after Albert Johnston bled to death from a wound incurred during the surprise attack at Shiloh. By the end of May, fearful that his army would be destroyed by either disease or Halleck's forces, Beauregard withdrew to Tupelo, a spot he "considered very healthy." Medical records show, however, that malaria shadowed the Army of Tennessee as it moved deeper into Mississippi. In June 1862, 5,756 of Beauregard's men, out of a mean strength of 40,675, were officially diagnosed with malaria. On average 80 out of every 1,000 soldiers suffered from "intermittents," and 61 out of every 1,000 were treated for "remittent fever." Again, the actual number of malaria infections was probably higher because the troops, most of whom had grown up in endemic areas, were forced to spend countless hours huddled together in relatively small areas where mosquitoes could easily spread parasites from one bloodstream to another.[3]

Having maneuvered the Confederates out of Corinth, Halleck marched his triumphant but sickly troops into the city. He knew Washington would want him to pursue Beauregard's army, but he was reluctant to send his men on what he believed would be a suicide mission. "If we follow the enemy into the swamps of Mississippi," he wrote Secretary of War Edwin Stanton, "there can be no doubt that our army will be disabled by disease." For the moment Halleck was content to send a "corps of observation" to keep an eye on the enemy. His decision not to pursue Beauregard's army was a product of his cautiousness. "Old Brains" preferred textbook operations and believed that marching a sick army through the unhealthiest part of the country at the unhealthiest time of year was not a move that Antoine-Henri Jomini would have endorsed. Like most nineteenth-century Americans, he viewed the southern disease environment as dangerous to outsiders. Proof was in the pallid and peaked faces of his men.[4]

Farther south, Butler was learning the same lesson. With yellow fever hanging over his head in New Orleans like the sword of Damocles, malaria and other diseases steadily eroded the health of his troops encamped in the surrounding countryside. The Fourteenth Maine Infantry pitched its tents near the swamps outside Carrollton, Louisiana, and in September and October 1862 "suffered terribly" from malaria. In less than a month the strength of the command dropped from seven hundred men fit for duty to a scant fifty-six. Butler noticed that Carrollton was especially unhealthy for his troops and was surprised when even a "regiment of acclimated Louisianans" he sent there was decimated by "malarious swamp fever." These natives, while perhaps accustomed to the local parasites, had no immunity to the new strains of malaria which were being brought in by men from all over the country. Their fellow Federals in the Twelfth Maine Infantry picked up plasmodium parasites in the summer of 1862 while on duty near Lake Pontchartrain and were afflicted by recurring bouts of intermittent fever well into the following winter. Camp Parapet, located seven miles north of New Orleans, quickly became known as an insalubrious location, in part because of malaria. The Eight New Hampshire Infantry was stationed there in June 1862 and suffered from typhoid fever, diarrhea, and "acclimating fever." Not surprisingly, the prevalence of malaria for the Department of the Gulf as a whole increased as mosquitoes became more active and the sickly season got under way.[5]

General Winfield Scott's warning early in the war about the dangers posed by the "malignant fevers" below Memphis was proving prescient for Butler's troops. But the worst was yet to come. A month after he forced New Orleans to capitulate, Farragut steamed up the Mississippi and ordered the surrender of Vicksburg, one of the last Confederate strongholds on the river. His fleet was rebuffed by Rebel regiments and cannon commanded by Martin Luther Smith, a Mexican War veteran and engineer who had once served under Robert E. Lee. The Lincoln administration, anxious to gain source-to-mouth control of the Mississippi River as quickly as possible, was displeased with Farragut's failure and pressed him to make a second attempt on the city. The sixty-year-old salt thought an attack on Mobile would be wiser but dutifully followed orders. In late June he took the fleet and three thousand soldiers under Brigadier General Thomas Williams toward the Confederacy's "Gibraltar of the West." Awaiting them were ten thousand Confederates and an untold number of anopheles mosquitoes. Both proved to be lethal deterrents.[6]

Before arriving on the twenty-fifth, Williams received orders from Butler to dig a four-by-five-foot ditch across the peninsula opposite Vicksburg with the hope of changing the course of the Mississippi River—a daunting engineering project that, if successful, would allow Union gunboats to bypass the city and gain control of the waterway. The chief result of this endeavor was the spread of disease among the U.S. soldiers and sailors who took part in it. Almost as soon as the Union fleet arrived at Vicksburg, malaria became a problem. The Mississippi River was ebbing after a recent flood and had left behind large puddles "covered with a thick green scum" which served as perfect breeding pools for the local mosquito population. Farragut's chief surgeon, Jonathan M. Foltz, commented upon the large number of these insects he saw on the way upriver but had no way of knowing they would soon cause his sick list to swell. By mid-July so many members of the Seventh Vermont Infantry were ill with "malarial diseases" that the regiment could barely muster enough men for guard duty. The same month the Second Massachusetts Light Artillery unit had only 21 out of its 140 men healthy enough to take part in a mandatory review and inspection, while their fellow soldiers in the Ninth Connecticut Infantry could not find enough troops to serve as funeral escorts for dead comrades.[7]

Mosquitoes, dirty water, and poor nutrition were responsible for the malaria, dengue fever, dysentery, and scurvy that broke out among Williams's regiments, but malaria quickly became the predominant disease at Vicksburg. The disorganization of the Union medical department in 1862 meant that most of the soldiers sent into Mississippi were undersupplied with quinine and ran out of the drug shortly after the campaign began. In at least one case, a request sent to headquarters for more of the antimalarial medicine was turned down on the grounds of "irregularity." When even the army hospital ran low on quinine and patients began to suffer from "new attacks and relapses," surgeons resorted to the use of cinchona bark, with successful results. The malaria problem was also exacerbated by the enlistment of several hundred black "contrabands" to assist in the digging of the canal. Having spent most, if not all, of their lives in an endemic environment, these native southerners were almost certainly carrying plasmodium parasites when they were recruited to work alongside Williams's nonimmune New Englanders. The men of the Thirtieth Massachusetts Infantry Regiment suffered from "a malignant form of remittent fever" after working on the canal, and by the time they left Vicksburg, more than half of them had to be placed on the sick list. Two hundred of the men went to the hospital, and an even greater number were unable to leave their quarters. A shortage of tents, which forced the regiment to sleep in jerry-built lean-tos, compounded the problem by giving the large number of insects in the area unrestricted access to the men's bodies at night. The canal project also involved felling trees and excavating earth, two activities that allow mosquito populations to multiply rapidly.[8]

By the last week of July, with the river falling faster than his sick men could dig, Williams had had enough. He wrote to Flag Officer Charles Davis, who had come down after a successful campaign in Memphis to help with the operation, explaining the reason for his withdrawal. "I will not attempt to disguise that I regret to leave this exceedingly," lamented the general. "I am really distressed to have effected so little at so much cost to my command in its effective strength, but with an increasing sick list, which must soon, if it has not already, reduced me below the ability for effective service, I have no alternative but to go." When Williams and Farragut abandoned the campaign at the end of July and retreated back down the Mississippi River to Baton Rouge, 75 percent of the 3,200 troops they had brought up from New Orleans were either dead or hospitalized, primarily because of malaria. Forty percent of Davis's men were also ill, prompting the naval commander to conclude that the North should "yield to the climate and postpone any further action at Vicksburg till the fever season is over."[9]

The Union withdrawal emboldened Confederate commanders. At the end of July Major General Edmund Kirby Smith sent a letter to his superior, Braxton Bragg, explaining why he thought the Gulf States were safe from any further attacks. "The enemy will, I think, attempt no invasion of Mississippi or Alabama this summer," Smith wrote. "The character of the country, the climate, and the necessity for concentration east are insurmountable obstacles." He went on to argue that there was still time "for a brilliant summer campaign." Bragg agreed and in August dispatched Smith and twenty-one thousand Confederates into Kentucky to begin a campaign that would culminate in the Battle of Perryville. In the wake of the Union retreat from Mississippi, the shaggy-browed general felt confident launching the invasion in the West which ultimately coincided with Lee's drive into Maryland.[10]

Although a number of factors accounted for Farragut's failure to capture Vicksburg—including strong Confederate defenses, the city's high altitude, and the rapid fall of the Mississippi River—anopheles mosquitoes also played an important role. The Union military could not afford to sustain such high disease casualties indefinitely and was forced to concede defeat. This loss, coupled with McClellan's failure to capture Richmond during the same period (a campaign also affected by malaria), helped convince the Lincoln administration that only the complete subjugation of the South, including the dismantling of slavery, would restore the Union and bring peace. When the president told his cabinet that summer that he had decided to make emancipation one of the North's primary war objectives, Secretary of State William Seward advised him to keep the decision under wraps until the Union achieved victory on the battlefield. Otherwise, reasoned Seward, the measure might look like the final act of a desperate and doomed government. Lincoln agreed and withheld his plan from the public until Union forces halted the Confederate invasion of Maryland in September 1862 at Antietam Creek.

The president's decision to emancipate slaves and arm them for the purpose of killing their former masters represented a radical departure from his earlier policies. Liberal Union generals such as David Hunter, James Lane, and John Phelps had already organized a handful of black regiments, but most of them were disbanded by a White House fearful of the political consequences. The military reversals of 1862, however, convinced Lincoln that emancipation and black enlistment were military necessities. Both policies strengthened northern forces while robbing the Confederacy of its chief labor force. Union medical personnel saw the issue from a slightly different perspective. A majority believed that African-American soldiers could better resist the diseases peculiar to the southern climate. This stereotype—a consequence of some blacks' limited immunity to mosquito-borne illness—lingered for the duration of the war, even though it was challenged by surgeons who had African-American patients directly under their care. "Notwithstanding the supposed exemption of negroes from the climatic diseases of the South, I am constantly seeing cases of the same fevers and diarrhœas among them, which prevail among the soldiers, and apparently of equal severity and frequency," wrote one northern surgeon from Memphis in August 1862, after treating a number of sick contrabands. "I am inclined to think that their power of resisting Southern climatic influences is greatly overestimated, though there is doubtless something to justify the common opinion on the subject."[11]

His enlightened views were not shared by most other Union officials. When Brigadier General John Phelps raised five companies of African-American soldiers in New Orleans without Washington's approval, Butler told his subordinate to put them to work clearing brush and fortifying the rugged terrain between Lake Pontchartrain and the Union lines. When Phelps threatened to resign rather than obey an order he thought degrading to Union soldiers, Butler gently reminded him that only the president could authorize the raising of black units, something Lincoln had not yet done. Butler, who would flip-flop on the issue in a matter of weeks, also seemed genuinely surprised that Phelps failed to recognize the wisdom of his instructions: "You have said the location is not healthy to the soldier. It is not to the negro. Is it not best that these unemployed Africans should do this matter at the present time?"[12]

Treasury secretary Salmon P. Chase, an early proponent of the use of African Americans as Federal soldiers, wrote to Butler in the wake of the Vicksburg fiasco explaining how the disease problem in the Gulf States and black freedom were related matters: "We cannot maintain the contest with the disadvantages of unacclimated troops and distant supplies against an enemy enabled to bring one-half the population under arms with the other half held to labor, with no cost except that of bare subsistence for the armed moiety." For Chase a federal emancipation scheme would not only swell the Union ranks but would do so with men who could better endure the disease-ridden swamps of Mississippi. General Williams knew firsthand the "disadvantages" to which Chase was referring. He noted that the summer of 1862 was an especially unhealthy season along the river and that even those troops not on the sick list were "yet so affected by malaria as to be good for nothing." Even the supposedly acclimated locals looked "thin, pale spiritless and yellow" to Williams. Several days after the withdrawal, he consoled himself with the hope that the Rebels at Vicksburg were equally sick.[13]

Major General Earl Van Dorn's Confederates were indeed ill, and malaria was responsible for a good deal of their suffering. W. J. Worsham was with the Nineteenth Tennessee Infantry and recalled that almost all the men in his outfit had "contracted chills" while stationed at Vicksburg. His description of the insect activity in the area explains why: "The enemy's shells annoyed us, but there was another foe we had to contend with, more annoying than the enemy's shells—the musquitoes, or, as the boys called them, 'gallinippers.' Roll up in your blanket ever so well, they would bite you. They would either get on the blanket with you and roll up with you, or they would bite you through all the folds." Worsham and the men of the Nineteenth wanted "to leave the land of flowers, magnolias and chills," but they were unable to obtain a transfer. A contingent of Kentuckians (William Preston's brigade) arrived in Vicksburg "in as good health as they had ever been since they entered the service" but quickly fell prey to the poisonous climate of the Mississippi Delta. For most of these men it was their first season in the Deep South, and like their Federal enemies, they were unaccustomed to its diseases. In June the brigade had 1,822 men fit for duty, but by July the number had dropped to just over 1,200. Even soldiers who had grown up in the region got sick. A group of Alabamians at Vicksburg was hit hard by disease, while "measles, malaria," and other ailments put close to 700 Louisianans out of commission in only one month's time.

Granville Alspaugh was a sixteen-year-old private with the Louisiana Twenty-seventh Infantry. "I tell you the chills & fever will hold you a long time if you dont do something for them," he wrote to his mother after suffering with paroxysms every other day for two months. During one such episode Alspaugh was sweating so profusely he "could hardly stand." The teenager also contracted measles but was eventually able to recover from both diseases. In spite of the widespread sickness in his command, Van Dorn decided that the time was right for an attack on the Union garrison at Baton Rouge and, if everything went according to plan, possibly on New Orleans as well. Williams's Federals occupied the Louisiana state capital and were still recovering from the malaria they had acquired at Vicksburg when Van Dorn set his attack in motion on July 27. The Confederate commander ordered former vice president

John C. Breckinridge to take five thousand healthy soldiers and connect with a smaller force stationed at a camp in northern Louisiana under the command of Massachusetts native Daniel Ruggles. From there Breckenridge and Ruggles were to march eighty miles south and capture Baton Rouge in order to "open the Mississippi, secure the navigation of Red River, then in a state of blockade, and also render easier the recapture of New Orleans."[14]

Malaria affected both armies during this campaign, which culminated in the Battle of Baton Rouge on August 5, 1862. Breckinridge could only find fewer than four thousand men healthy enough to make the trip from Vicksburg to Camp Moore (a "good number" could not go because of "chills"), and by the time the battle actually got under way, he had only twenty-six hundred effectives, with the rest either sick or dead. In one twenty-four-hour period, which included a "night march," he lost four hundred men to disease. The general blamed "the effects of exposure at Vicksburg" for the "appalling" sick rate among his men, who were poorly clothed (many without shoes or coats) and had spent the hottest part of the summer at Vicksburg deprived of regular doses of quinine. Many of these soldiers were already carrying plasmodium parasites when they arrived in Louisiana and were experiencing recrudescent fevers as they marched toward Baton Rouge. The intense thirst generated by these fevers may explain why so many men ignored the advice of their surgeons and drank from the stagnant puddles they passed on the way to the front, causing an outbreak of dysentery. Some soldiers also ate "green corn" straight out of the fields during the march which made them sicker still.

Williams's bluecoats were in bad shape as well. He supposedly had seven thousand troops with which to defend Baton Rouge but could only find twenty-five hundred men healthy enough to fight on the day of the battle. Half of these men were still recovering from "malaria and chronic diarrhea" and were too weak to dig trenches around the city in preparation for the Confederate attack. As Breckinridge's forces swung into action on the morning of the fifth, men on both sides were fighting sickness as well as each other. Confederate lieutenant L. L. Etter was suffering from a "high fever and intense thirst" and was shaking so violently he could barely stand but managed to stay with his fellow Tennesseans and battle the Fourteenth Maine Infantry, which was itself not at full strength on account of disease. A number of Union soldiers got up out of their sickbeds to participate in the battle. One captain with the Seventh Vermont Infantry, who was "seriously ill," made a heroic effort to get to his unit, but exhaustion kept him from making it before the battle ended. S. K. Towle was a surgeon with the Thirtieth Massachusetts Infantry at Baton Rouge and reported that a minimum of four hundred of his men had been ill each day during the month of August "almost entirely from malarial diseases, chiefly remittent fevers."[15]

After a five-hour fight Breckinridge pulled his troops back to the Comite River, when word came that the CSS *Arkansas*—the South's ironclad answer to the Union gunboats supporting Williams's forces—had broken down four miles away. The exhausted and disorganized Union soldiers were in no condition to give chase, having lost their general to a Rebel musket ball earlier in the day. When Van Dorn learned of the Confederate repulse at Baton Rouge, he moved Breckinridge's forces to Port Hudson, a key location on the Mississippi with greater strategic value than the Louisiana capital. With both Vicksburg and Port Hudson occupied by men wearing gray uniforms, Richmond controlled a hundred-mile stretch of the Mississippi River and was able to keep both halves of the Confederacy connected for another year. Seeing the futility of occupying a city with no real military value and worried about the safety of New Orleans, Butler withdrew his soldiers to the Crescent City. He praised them for having "repulsed in the open field" an enemy "who took advantage of your sickness from the malaria of the marshes of Vicksburg to make a cowardly attack."[16]

To the large number of men who were ill, Butler's words offered little comfort. They arrived in New Orleans in a condition that shocked surgeon Eugene Sanger. "The patients complained of burning in the stomach and exhaustion," he later reported. "They seemed wholly unconcerned whether they lived or died, and continually tossed to and fro until death relieved them from their sufferings." A number of these cases responded to quinine treatments, indicating that malaria was definitely present and, based on the symptoms, may have been *Plasmodium falciparum*.

Two months after being dispatched to capture Vicksburg, these regiments were back where they had started, a shadow of the proud and optimistic units they had once been. Disease, chiefly malaria, had helped check the campaign against Vicksburg and created a stalemate in Baton Rouge which ultimately gave the Confederacy an advantage in the region. If either side had been at full strength on August 5, it might have captured or destroyed the other and reigned supreme in lower Louisiana. Instead, Butler retreated to the relative safety of New Orleans, while Van Dorn secured Port Hudson, which was almost immediately reinforced by a Louisiana artillery battalion that authorities in Richmond thought "could stand the climate." The Mississippi River remained closed until another general with greater resources and ambition found a way to take Vicksburg out of the war.[17]

While mosquitoes sickened Butler's and Halleck's armies out West, the insects were adversely affecting Union military operations in the East. Four months before yellow fever killed Ormsby Mitchel, malaria had helped convince his predecessor, Major General David Hunter, to abort a campaign to capture Charleston. In early June 1862 Hunter had landed two divisions on the southern end of James Island (just below Charleston) and planned to move his forces up the Stono River toward the city that was the birthplace of secession. Capturing Charleston would have dealt a severe blow to southern morale and provided a launching point for future Federal operations. Facing a determined band of Confederates under the command of Brigadier General Nathan G. Evans, Hunter decided to delay his advance and wait for reinforcements.

As the Union troops performed picket duty and awaited orders, clouds of mosquitoes attacked them "night and day," leaving behind "poisonous and irritating" bites. One Connecticut soldier thought the insects "were the most awful enemies he ever encountered." He and his battery camped in a cotton field and were "nearly devoured by fleas and mosquitoes." Hunter's careful planning, however, included daily drams of quinine and whiskey, which controlled the malarial parasites swimming in the soldiers' bodies. But enforcement was lax, and many soldiers chose not to take the bitter-tasting medicine, especially after surgeons started dispensing it without whiskey. Malaria soon began to take its toll, along with other diseases such as typhoid fever. When one of Hunter's subordinates, Brigadier General Henry Benham, disobeyed orders and launched a disastrous offensive that culminated in a Union loss at the Battle of Secessionville, Hunter realized his position was untenable and ordered a withdrawal. He wrote to Brigadier General H. G. Wright (Benham's replacement) that having heard "from Washington that there is no probability of our receiving re-enforcements, and it being all-important to provide for the health of the command in the sickly season approaching, I have determined to abandon James Island, in order that the troops may be placed where, in so far as practicable in this climate, they may be out of the way of malarious influences, and where the picket duty will not be so exhausting on our men as at present."[18]

Hunter knew that Washington was focused on the campaigns in Virginia and that the soldiers under his command would be decimated by disease if they dallied too long on James Island—a location reputed to be unhealthy even among supposedly acclimated southerners. His men complained about "the miasma rising from the saturated ground" and the suffocating vapors they believed were caused by a combination of heavy rains and broiling heat. Their Confederate enemies on James Island also suffered from malaria. Even before Hunter's forces landed, South Carolina governor Francis Pickens worried that Charleston might be vulnerable to Union attack, knowing that a southern defense force could not survive on the island during the sickly season. John G. Pressley and a contingent of South Carolina volunteers guarded the area after Hunter's forces were withdrawn and noticed that the mosquitoes were "fearfully troublesome." By July malaria was "very common among the troops," and quinine was being offered on a regular basis by Confederate surgeons. But southern troops, like their northern counterparts, often disregarded their surgeons' instructions and chose not to take the bitter-tasting medicine. In addition, the Union blockade was causing price hikes and shortages of the drug in the South which made it difficult to dole out effective doses consistently. As a result, Pressley's sick list expanded until an average of "seventy to one hundred" soldiers a day fell ill. An outbreak of typhoid fever compounded the problem, and his men suffered until the first frosts appeared in October and halted the activity of the mosquitoes.[19]

Union troops stationed further north, in eastern North Carolina, were also battling mosquito-borne illness. The previous year a joint army-navy operation led by Butler had secured a northern beachhead on Carolina's Outer Banks. Over the next six months U.S. forces captured nearly every important coastal port in the "Old North State" (Wilmington, a notable exception, remained in Confederate hands for most of the war).

But North Carolina's sickly season, which was sicklier than most, taxed the health of the occupation force that was subsequently sent there. In a single year the U.S. military recorded 2,353 cases of malaria "per 1,000 mean strength," leading the Department of North Carolina to earn the ignoble distinction of being the second "most malarious" district in the South. Only Arkansas was worse. Troops stationed in Washington and Plymouth and along the rail line between New Bern and Morehead City suffered severely from "chills," "congestive fever," and "intermittent fever," especially during the autumn months. Picket duty and reconnaissance missions sickened "nearly every man" in the Seventeenth Regiment Massachusetts Volunteers, which reported 1,100 cases of malaria in one month alone. The regiment was taken off picket duty, housed in New Bern, and successfully treated with quinine. Quinine was irregularly administered to the soldiers of the Pennsylvania Fifty-first, whose ranks were soon decimated by "diarrhœa, dysentery, ague, camp fever, and small-pox." By April 1862 only 350 of the original 981 men in the regiment were able to fight. Two Connecticut regiments stationed near New Bern were in a similar predicament. The principal problems diagnosed among the men by their assistant surgeon were "malaria, diarrhoea, and malingering." As the War Department continued to send fresh troops into the area, new infections multiplied. Surgeon D. W. Hand took over as medical direc-tor for the Union forces in North Carolina in August 1863 and was swamped with innumerable cases of intermittent and remittent fever over the next three months. According to Hand, between 40 and 168 men per regiment stationed near New Bern were sick each day, and several died of congestive fever. Most of these men likely contracted malaria while on patrol in the North Carolina countryside since the disease was not a problem within New Bern's city limits.[20]

These events largely fulfilled the expectations of Confederate com-manders, who were confident that southern diseases would stymie Union thrusts into the Carolina interior during the sickly season and planned accordingly. As Major General George B. McClellan's army moved sluggishly toward Richmond in the late spring of 1862, Robert E. Lee sent a letter to commander of the Department of South Carolina and Georgia, John C. Pemberton, requesting reinforcements. "At this season I think it impossible for the enemy to make any expedition into the interior," wrote Lee. "The troops that you retain there will suffer more from disease than the enemy." Lee's bleak assessment was not far from the truth, but when Hunter's troops landed on James Island in June, Pemberton sent back a panicked plea for more guns. The reply came not from Lee but from Jefferson Davis, who also requested rein-forcements from Pemberton to halt McClellan's advance on Richmond.

"Decisive operations are pending here in this section," he wrote, "and the climate already restrains operations on the coast." Two days later Davis wrote to Pemberton again, asking for at least three regiments: "It was hoped the season would secure you against operations inland, and that you could spare troops without weakening your strength for the defense of Charleston." Davis, who knew the hazards of the southern disease environment as well as anyone and trusted Lee's advice, explained his rationale more carefully in a letter to South Carolina governor Francis W. Pickens. Pickens was concerned that Charleston's security might be compromised if Pemberton's troops were transferred. Davis clarified that he only wanted soldiers "from positions where the season will prevent active operations," which included the mosquito-infested areas along the Charleston & Savannah Railroad such as Coosawhatchie and Pocotaligo. Months earlier Lee, then commanding the department, had posted Confederate troops at these spots as part of his overall defense strategy for the region. Realizing that Southern cannon were no match for Union gunboats, the Virginia general had pulled Rebel regiments back from the coast in order to guard the railroad line that allowed him to move men and materiel rapidly from one threatened front to another. Pemberton eventually got the point. When he became convinced that Hunter was not an immediate threat, he sent four regiments to Virginia that helped check McClellan's advance and save the Confederate capital from capture.[21]

A snapshot of the Union disease casualties among the men of the Department of the South—a command that included South Carolina, Georgia, and eastern Florida—in 1862 reveals why Davis and Lee were confident they could transfer troops from the Carolina theater to other parts of the Confederacy. Between April and November northern surgeons diagnosed 3,375 cases of intermittent fever in addition to 2,690 cases of remittent fever. Over 6,000 cases of diarrhea and dysentery—the chief medical ailments plaguing the department—were diagnosed during the same period. Catarrh, mumps, typhoid fever, and rheumatism added to the soldiers' misery. But even these carefully compiled statistics probably offer a low estimate of the prevalence of malaria in the department. Soldiers who feared hospitals routinely avoided treatment and likely suffered through periodic bouts of "ague" without complaining. Quinine could not eliminate plasmodium parasites from the bloodstream, and they could become active again later. Because troops all along the southeastern seaboard noticed the staggering number of mosquitoes swarming in their midst, it is highly probable that many of those diagnosed with other maladies were suffering from malaria at the same time, a condition known as disease "synergism" in the modern medical vernacular.[22]

Confederates stationed in the Department of South Carolina, Georgia, and Florida also suffered extensively from malaria. Between January 1862 and July 1863 the department had a mean strength of 25,000 troops but recorded more than 41,000 cases of the disease. In Georgia mosquitoes hatched in the flooded rice fields lying along the Savannah River swarmed nearby Rebel garrisons, producing 3,313 cases of malaria among an average of 878 men in only fourteen months. The situation was similar for troops stationed across the border in South Carolina. Out of 5,735 diagnoses made at the Wayside Hospital in Charleston during one eight-month period, 1,535 were for malaria. Confederate surgeons' experiments with quinine to determine whether or not it could be used as a prophylactic provide a good indication of how prevalent the disease was in South Carolina. Out of 506 soldiers administered quinine on a regular basis, 98 fell ill, while 134 of the 234 "who took no quinine" at all contracted a "fever."[23]

While South Carolina's malaria problem bolstered the confidence of Confederate planners in Richmond, it intimidated the Union commanders operating in the region. When Mitchel replaced Hunter as commander of the Department of the South in September 1862, he too allowed fear of southern diseases to affect his plans in the Palmetto State. In the weeks before he was killed by an *Aedes aegypti* mosquito, the general cut short a raid near Pocotaligo—an operation designed to sever the Charleston & Savannah Railroad—in part out of concern for the health of his troops in the interior. In a letter to a friend Mitchel shared the reasons for the abrupt departure, arguing that he could not "yet make a lodgment on the mainland, as the country fever would probably kill off all my troops." He echoed this concern in an official report he subsequently sent to Washington at the end of October. The general's uncharacteristic cautiousness confirmed the accuracy of Lee's and Davis's predictions and allowed the Charleston & Savannah Railroad to remain operational. By retaining control of this important line and others in the Carolinas for most of the war, the South was able to deploy forces rapidly to threatened points on the coast and keep its eastern armies supplied with goods smuggled in from Europe and the West Indies. In short, mosquitoes were helping the Confederacy retain control over its interior lines (a key component of the slaveholding nation's defensive strategy) while simultaneously shielding its shorelines. In the fall of 1862 Mitchel found himself caught in this entomological pincer move; in the Carolina countryside malaria posed a grave threat to the health of his unacclimated soldiers, while yellow fever endangered his position along the coast at Hilton Head.[24]

The same was true for the Union military as a whole during the 1862 sickly season. *Aedes aegypti* mosquitoes swarmed the Yankee sailors and soldiers stationed along the coastline between South Carolina and Texas, while anopheles mosquitoes steadily eroded the health of their comrades who were ordered inland. Northern commanders, aware of the poor health of their men thanks to regular reports from their medical officers, were reluctant to launch operations during the unhealthiest months of the year. Halleck, for example, refused to pursue Beauregard into Mississippi in part because of concerns over what southern diseases might do to his men. It may be tempting to dismiss this cautiousness as little more than a convenient excuse for inaction until we reflect upon what happened to the Union forces that actually entered the bowels of the South during the fever months. The North's plan to build a canal around Vicksburg during the summer of 1862—part of a bold first attempt to capture that city—collapsed when more than half of the "non-acclimated" soldiers succumbed to various diseases, mainly malaria. Halleck, who had proven himself a fighter, was not about to let a similar scenario destroy his army. He took these concerns with him to Washington in July 1862, when he was promoted to general-in-chief of all the United States's armies and turned his attention to the operations of his new subordinate, Major General George B. McClellan. McClellan at the time was busy leading a colossal Union force through the swamps of eastern Virginia, another area of the South rife with mosquitoes and disease.

HARD TIMES IN OLE VARGINNY, AN' WORSE A CUMIN'!
Scene.—Rebel Pickets in Western Virginia.

FIRST PICKET. "Awful Cold, ain't it?"
SECOND PICKET. "Co-o-ld! yes, an' I'm jist gitting another Shake of that Ager, and no Quinine in the 'Federacy!"
FIRST PICKET. "Worser still! Got them Blue Devils after me, an' nary drop o' Whiskey." (*With much feeling.*)
SECOND PICKET. "I wish I was Ho-o-me."
[*They part, singing, mournfully,* DIXIE, *without the Variations.*]

The South's quinine shortage contributed to the poor health of Confederate soldiers. Recrudescent fevers often occurred months after an initial plasmodium infection, as this 1862 *Harper's Weekly* cartoon shows. *Source: Harper's Weekly,* 6, no. 262, January 14, 1862, 16. (Courtesy Collection of the New-York Historical Society, negative no. 81468d)

Although nineteenth-century physicians did not understand the link between mosquitoes and disease, the use of nets helped troops stay healthy. Here patients at Harewood Hospital in Washington, D.C., pose for a photograph. (Library of Congress, LC-B815-1008)

Benjamin Butler's rigid quarantine and sanitation measures helped prevent a yellow fever epidemic from occurring in New Orleans during the Union occupation. Here a cartoonist depicts President Lincoln graciously receiving his general after a job well done. *Harper's Weekly* 7, no. 316, January 17, 1863, 48. (Library of Congress)

Major General Ormsby MacKnight Mitchel died of yellow fever while commanding the Union Department of the South. (Courtesy U.S. Army Military History Institute, folder RG100S–George L. Febiger Coll.31b)

General Thomas Williams's attempt to dig a canal opposite Vicksburg during the summer of 1862 produced a severe outbreak of malaria among his men, which helped scuttle the North's plan to gain quick control of the Mississippi River. Here contraband slaves, many of whom were likely carrying plasmodium parasites, assist with the project. *Harper's Weekly* 6, no. 292, August 2, 1862, 481. (Courtesy HarpWeek, LLC)

Luke Pryor Blackburn, M.D. Dr. Blackburn's attempt to start a yellow fever epidemic inside the nation's capital was not enough to keep him from winning the 1879 Kentucky gubernatorial race. (Courtesy Kentucky Historical Society, accession no. 1904.4)

ADVANTAGE OF "FAMINE PRICES."

Sick Boy. "I know one thing—I wish I was in Dixie."
Nurse. "And why do you wish you was in Dixie, you wicked boy?"
Sick Boy. "Because I read that quinine is worth one hundred and fifty dollars an ounce there; and if it was that here you wouldn't pitch it into me so!"

The effectiveness of the Union blockade increased the price and scarcity of quinine in the Confederacy, causing additional hardships for the southern civilian population. In this 1863 *Harper's Weekly* cartoon an ill northern child wishes he lived in an area with less of the bitter-tasting medicine. *Harper's Weekly* 7, no. 359, November 14, 1863, 736. (Courtesy HarpWeek, LLC)

BEFORE PETERSBURG—ISSUING RATIONS OF WHISKY AND QUININE.—[SKETCHED BY A. W. WARREN.]

By 1864 the North's superior supplies of quinine helped Union troops stay healthier than their Confederate foes. Here federal soldiers at Petersburg receive a prophylactic dose of the drug, which was often mixed with whiskey to make it more appealing. *Harper's Weekly* 9, no. 428, March 11, 1865, 145. (Courtesy HarpWeek, LLC)

{5}

"The Pestilent Marshes of the Peninsula"

MALARIA WAS ONE of many problems that beset Major General George B. McClellan's army as it trudged through the brackish swamps of Tidewater Virginia en route to Richmond. "Little Mac" was confident that his plan to capture the Confederate capital by pipelining 100,000 Federals up the long peninsula formed by the York and James rivers would end in a smashing victory for the North and a return to the antebellum status quo. Lincoln was less certain about the soundness of McClellan's plan but, lacking a better option, gave his general the troops and transports he requested. McClellan's target, coastal Virginia, had been known as a hotbed of malaria since colonial times, and the influx of huge numbers of men and animals during the Civil War provided a plethora of new prey for its large mosquito population. At Yorktown a year before Union troops showed up, Brigadier General John Bankhead Magruder's Confederates suffered from "ague" much as Lord Cornwallis's army had on the same spot eighty years earlier. When his supply of quinine ran low at one point, Magruder informed Richmond that the medicine was "absolutely necessary" to shrink the sick list, which had grown "frightful" without it. The Army of the Potomac would learn the same lesson when it arrived in the area, especially after heavy rains in late May and early June 1862 created innumerable breeding pools for anopheles mosquitoes, which thrived as a result of warm summer temperatures and a superabundance of food.[1]

These mosquitoes caused widespread suffering among the unacclimated Union regiments that took part in the Peninsular Campaign. Regis De Trobriand was a blue-blooded French immigrant who at the time had command of the Twelfth New York Light Artillery. During the march toward Richmond he and his men spent the night on President John Tyler's plantation near Charles City, next to "a thicket which was interspersed with stagnant pools." The next morning Trobriand experienced "aguish feelings," which grew worse until he was forced to retire from the campaign. In later years he remembered the pestilential Virginia climate as being more dangerous to his men than the Confederate army. A surgeon attached to one of the Twelfth's sister regiments, the Fifth New York Infantry, believed June was the sickliest month of the whole campaign. He noticed that the men assigned to picket duty at night were especially susceptible to disease, as they were "exposed to miasm in its most concentrated form." Other regiments from northeastern states also suffered in the Virginia climate. The Massachusetts Thirty-second Infantry Regiment came straight "from the sea air of New England" and was quickly laid low by malaria. Men filtered in and out of the hospital for weeks, making the regiment unfit even for drill. Their comrades in the Fifth New Hampshire Infantry were camped in a swamp near the Chickahominy River which was full of mosquitoes, "moccasins and malaria," reported the regiment's historian, who also served as its surgeon for a time. He believed the "exhaustion, suffering, disease, and death" suffered by the men of the Fifth beggared description.[2]

Malaria plagued the Union army during the Peninsular Campaign for a number of reasons. First, the U.S. military's medical department was still disorganized in 1862. The mismanagement of the ambulance service and the autonomy of regimental surgeons meant that many sick and wounded soldiers went without treatment for long periods of time. The Army of the Potomac's medical director, Charles Tripler (who would soon be replaced by the more adroit Jonathan Letterman), insisted on keeping the sick with the army rather than evacuating them, which meant plasmodium parasites were more easily passed to noninfected troops in crowded hospitals. Compounding the problem was the army's supply nightmare, which prevented adequate amounts of quinine from reaching regimental surgeons. Dr. William Smith was a surgeon serving with the Eighty-fifth New York Infantry who watched malaria and diarrhea sicken nearly half the men in his regiment. Unable to obtain quinine during the campaign, Smith suspected the scarcity of the medicine was to blame for the army's high rate of sickness. His suspicions were confirmed weeks later when he obtained a supply of the drug and "quickly restored the fever cases" in his unit. Tripler was so exasperated by the lack of quinine that he pleaded for the medicine in a letter to the surgeon general on May 29, two days before the Battle of Seven Pines began. As a result of these shortages, quinine was used only sporadically as a prophylactic, and troops quickly fell ill. Also, many of the soldiers in McClellan's army had spent a considerable amount of time training in the malarial flats around Washington, D.C., before the campaign began and were already carrying plasmodium parasites when they stepped off the transport boats at Fortress Monroe and were swarmed by hungry insects. Finally, McClellan chose to launch his offensive against Richmond through what Confederate secretary of war George Randolph called "the pestilent marshes of the Peninsula" at a time of year when anopheles mosquitoes were highly active. Had he moved more quickly or begun his campaign earlier in the year, it is unlikely that his army would have suffered as many disease casualties as it did.[3]

The biggest medical problems for the Army of the Potomac during the Peninsular Campaign were diarrhea and dysentery. In July alone, more than nineteen thousand Union troops were suffering from what they jokingly referred to as the "Virginia Quickstep." But because nearly eight thousand cases of intermittent fever were diagnosed the same month, it is likely that at least some of the soldiers suffering with dysentery got it after drinking dirty water from the Virginia bogs in order to slake a malaria-induced thirst. *Plasmodium vivax* alone was rarely lethal, but together with the other maladies that were besetting Union troops at the same time (typhoid fever and scurvy were also problems), its presence could determine whether a soldier lived or died. Dr. Joseph J. Woodward was an assistant surgeon with the Army of the Potomac who recognized that many of his patients were suffering from more than one disease at the same time. With the surgeon general's approval, he went on to create the category "typho-malarial fever," which was the diagnosis for nearly fifty thousand patients before the war ended. How many of these cases were actually caused by plasmodium parasites remains a mystery because misdiagnoses were frequent and the symptoms of typhoid fever are sometimes similar to those of malaria.[4]

· As the exhausted Army of the Potomac retreated to Harrison's Landing in July 1862 after a series of bloody engagements with Lee's forces, malaria and mosquitoes followed. The sick list was initially between 20 and 40 percent, but Letterman, the army's newly appointed medical director, thought the health of the troops was improving as the summer wore on. Brigadier General Erasmus D. Keyes, commander of McClellan's Fourth Corps, disagreed. He sent a letter to Lincoln urging the president to withhold reinforcements and withdraw the army before the sickly season destroyed it. "To bring troops freshly raised at the North to the country in the months of July, August, and September would be to cast our resources into the sea," he wrote. "The raw troops would melt away and be ruined forever." Keyes knew firsthand about fresh Union troops melting away. One of his raw divisions occupied "the worst [position] on the Peninsula" during the siege of Yorktown and had been hit hard by "malarial and typhoid fevers." Eleven days after he wrote to Lincoln, Keyes fired off another missive, this time to Montgomery Meigs, imploring the quartermaster general to use his influence with the administration to press for a withdrawal. His protests did not fall on deaf ears. Although McClellan wanted reinforcements for another crack at capturing the Confederate capital, Lincoln decided that the campaign had gone on long enough. The president was weary of Little Mac's excuses, and his troops were needed in northern Virginia to fend off Thomas "Stonewall" Jackson's forces, which had been spotted at Gordonsville. In August Halleck sent a letter to McClellan explaining another important reason for the Union pullout. "To keep your army in its present position until it could be so re-enforced," the general-in-chief chided, "would almost destroy it in that climate. The months of August and September are almost fatal to whites who live on that part of James River." Like many Americans of his time, Halleck feared the pernicious diseases of the South's sickly season, which was in large part a result of mosquito activity. His concerns about the diseases at Harrison's Landing echoed those of his subordinates, Davis, French, Hunter, Mitchel, Renshaw, and others who were equally worried about the unhealthiness of the areas in which they were operating. All along the South's riverbanks and shorelines, anopheles and *Aedes aegypti* mosquitoes created biological hot zones that complicated the plans of Union commanders and damaged the health of their troops.[5]

Halleck's assessment was consistent with the one he had made at Corinth two months earlier as commander of the combined Union armies in the West. During the Peninsular Campaign "Old Brains" Halleck was again considering every angle, including the illnesses that threatened McClellan's troops, and determined that discretion was the better part of valor. His comments also make it clear that malaria and yellow fever were significant contributors to Tidewater Virginia's sickly season, given that some blacks enjoyed limited immunity to both diseases.[6] African Americans were just as susceptible as whites to typhoid fever, dysentery, and the other diseases that made life miserable for the Union troops crowded together at Harrison's Landing. Anopheles and *Aedes aegypti* had assisted in creating an image of Virginia as an unhealthy region before the war, and the high rate of malaria during the Peninsular Campaign helped expedite the Army of the Potomac's retreat to Washington. McClellan's defeat, in part attributable to disease, triggered a sea change in the North's approach to the war; afterwards it would work to destroy slavery and create a new birth of freedom rather than fight exclusively to preserve the old republic.[7]

The impact that malaria had on the southern armies that participated in the Peninsular Campaign is less clear. Confederate records on disease casualties are sketchy, but given that one out of every seven sick rebels stationed east of the Mississippi River in 1861–62 was diagnosed with malaria, it is reasonable to assume that the disease also caused considerable misery among the gray armies guarding Richmond. Herbert Nash, an assistant surgeon attached to several Virginia units, remembered that many "unacclimated confederates" suffered the same diseases as McClellan's troops after "traversing the marshes and slashes of the Chickahominy." A glance at the statistics compiled by individual Confederate hospitals provides some indication of the problem. Out of 366 cases recorded by Hospital No. 21 in June 1862, there were eighteen cases of remittent fever, seven cases of intermittent fever, and three cases of congestive fever. "Febris" cases, many of which could have been malaria, were not recorded. Hospital No. 18 kept records on 182 patients in August 1862 and saw 5 cases of remittent fever and 11 cases of intermittent fever but did not keep any information on general "fever" cases. But these statistics only include those soldiers who chose to seek treatment. Civil War soldiers usually avoided hospitals whenever possible. And unlike their enemies from New England, most southern boys had grown up in malarial environments and were accustomed to "the chills" to one degree or another. Furthermore, the Confederacy's supply of quinine was not yet dangerously low in early 1862, and a soldier infected with plasmodium still had a good chance of receiving effective treatment from his regimental surgeon. Although it remained a problem, malaria was less of an issue for the South's armies during the Peninsular Campaign because of access to ample supplies of medicine and provisions in Richmond and southerners' long experience with the disease.[8]

The situation was dramatically different a year later for the Confederates who were tasked with defending Vicksburg from Ulysses Grant's forces. Grant spent the winter of 1862–63 spinning his wheels in the mud (both literally and figuratively), trying to figure out how to take the Confederacy's "Gibraltar of the West" out of the war. Rebel cannon protected the river approach to the city, while the surrounding countryside was fortified by thickets of half-submerged trees and alligator-infested bogs. Early in the campaign Grant put his men to work on the canal that Thomas Williams had abandoned a year earlier. Sherman's troops dug through ankle-deep mud, but the unpredictable ebb-and-flow cycles of the Mississippi River turned the task into another Union exercise in futility. Mississippi's cooler winter temperatures halted the activity of mosquitoes during this period, but diseases such as typhoid fever and diarrhea were rampant. Undeterred by the canal failure, Grant explored several other water routes through the swamps of Louisiana and Mississippi before settling on a plan that showed the world that he possessed the heart of a gambler.

Grant's decision to run the Union fleet past Vicksburg's guns and march an army through the heart of the South without a proper supply line has long been viewed as a bold and brilliant masterstroke. What is often overlooked yet equally remarkable about the 1863 campaign (and speaks volumes about Grant's leadership) is that it took place during Mississippi's sickly season. Other generals were unnerved by the prospect of exposing their men to the South's dangerous diseases during the summer months. Even Robert E. Lee—arguably the greatest military tactician either side produced during the war—thought it highly unlikely that the North would march an army through Mississippi at the hottest time of year. The northern press expressed the same sentiments and argued that a siege against Vicksburg during the sickly months would be suicidal. "The simple substance of it is, that an army of twenty-five thousand men would find their graves between now and the first of October without ever facing an enemy," opined a correspondent for the *Chicago Times*. Even if Grant had chosen to ignore the opinions of the press and his superiors, he could not overlook the appalling amount of sickness among the troops under Williams's command during their failed campaign against Vicksburg.[9]

Fortunately for Lincoln and the Union war effort, Grant did not let these concerns stop him from launching a successful siege against Vicksburg in the summer of 1863. Again, Grant's success was in part attributable to his decision to abandon his supply lines and live off the fat of the land—a lesson he had picked up the previous winter, when Earl Van Dorn's cavalry had ridden around his army during its march against Vicksburg and destroyed the Union supply hub at Holly Springs. What is less well known is that medicine, including quinine, was not one of the articles Grant was willing to part with when he moved into Mississippi. The Federal host that menaced General Joseph E. Johnston's and General John C. Pemberton's armies was well stocked with pharmaceuticals and prepared for every conceivable medical emergency. In contrast, their opponents suffered severe supply shortages, which allowed malaria to become pervasive among the Confederate ranks. These deficiencies allowed the anopheles that had inadvertently served as Vicksburg's guardians a year earlier to become the city's enemies. Grant's supplies of medicine were so abundant, in fact, that he could occasionally afford to show mercy toward his enemies. After the Battle of Champion's Hill—the last major clash between the Yankees and Rebels before the siege of Vicksburg began—he made sure desperate southern doctors were assisted by his medical staff and given whatever supplies they needed. Dr. Thomas Barton was one of the Union physicians who treated Confederate stragglers for intermittent fever.[10]

Even with its large stockpile of medicines, however, the Army of the Tennessee was hit hard by disease as it marched through the Mississippi swamps. Diarrhea and malaria were the chief scourges, and roughly a quarter of Grant's forces were sick or incapacitated during the campaign. As early as March, one surgeon reported that the disease rate exceeded "17 per cent" for the troops operating in the vicinity of Vicksburg and that the sickest soldiers were those "camped near the canal" that Williams had started (and Grant failed to finish), where the terrain was "low and partially inundated." John Sherman Emerson was an assistant surgeon with the Ninth New Hampshire Infantry who watched the "climate and forced marching" wreck the health of the men under his care. By the time the Ninth was transferred to Kentucky, almost every soldier in the regiment had contracted malaria. Emerson's colleague, George Twitchell, was surgeon-in-chief for two divisions and treated numerous patients whose suffering was "greatly augmented by the fever and malaria." "Fevers, congestive chills, diarrhœa, and other diseases" made life so miserable for the men of the Fiftieth Pennsylvania Infantry that many simply "sank down upon the roadside" and died. During the siege of Vicksburg, A. J. Withrow wrote to his wife and explained what conditions were like in camp for the Twenty-fifth Iowa Infantry: "I did not get my letter finished the other day, and yesterday I came near shaking my toe nails off, with the ague, and did not feel like writing." His only solace was that the chills only came "every other day" instead of daily. Jacob Ritner was also with the Twenty-fifth but was lucky enough to stay healthy, unlike the other men in his unit who were afflicted by "mostly 'ague' and diarrhea." Despite the widespread sickness, Ritner was optimistic that the campaign would soon end. "Our situation here is very unhealthy, but I believe we will get out of it before too long," he wrote. Malaria-carrying mosquitoes would continue to poison the Twenty-fifth Iowa and the Army of the Tennessee's other regiments for another month. Union commanders were also worried about the possibility of a yellow fever outbreak. Orders went out from headquarters for regiments to avoid the "scorching rays of the sun" and "damp and chilly night air" whenever possible as a preventive measure against the disease. Fortunately for Grant, yellow fever never became a problem during the Vicksburg campaign.[11]

The hardships that malaria caused for the Union army paled in comparison to the misery it produced among Confederates. A third of Pemberton's force was already ill when the campaign began, and by July the number had climbed to nearly 50 percent. Like the Union army, diarrhea and malaria were responsible for most of the cases of illness, and a large number of the sick were undoubtedly suffering from both diseases at once. Men parched by the broiling Delta heat were forced to drink contaminated water while hungry insects buzzed in their ears day and night. Soldiers on picket duty were especially vulnerable to mosquito bites because they were not allowed to build fires, which might reveal their positions to enemy mortars. One lieutenant with the Thirty-fifth Alabama Infantry who was often assigned to picket duty (and remembered Vicksburg as "a place of flux and mosquitoes") was bitten so frequently that he found it difficult to sleep when not on watch. A Tennessee soldier named Samuel Swan was also tortured by mosquitoes at night and made numerous entries in his diary about the subject. The parasites spread by these mosquitoes coupled with inadequate supplies of food, clothing, and medicine reduced the Confederates at Vicksburg to an army of scarecrows. The hospitals were soon overflowing with patients. One eyewitness described the Rebels he saw at Vicksburg as sallow-cheeked and sunken-eyed, with an overall appearance that was "haggard & care-worn."[12]

A number of these men had only recently been transferred to Pemberton's army from South Carolina. During the summer of 1863 the Confederate high command in Richmond was again convinced that the sickly season would help protect the Carolina coastline. Coming off his stunning victory at Chancellorsville and planning an offensive that would culminate in the Battle of Gettysburg, Lee pushed Confederate secretary of war James Seddon for a troop transfer from the "southern coast" to the Confederacy's "northern and western frontiers." Lee thought that a local, acclimatized force could endure the summer months and was sufficient to take care of any Union recon units foolish enough to go traipsing through the South Carolina countryside in the middle of summer. Seddon did not need convincing and had already been transferring some of P.G.T. Beauregard's troops from there to Mississippi to help stop Grant's advance. When he received complaints about the transfers from South Carolina officials concerned about the safety of Charleston, Seddon assured them that the "sickly season" and "sultry weather" would help keep the city safe. Beauregard conceded that he could manage with fewer troops during the summer months.[13]

As Beauregard's men moved west to keep the Stars and Bars flying over Vicksburg, black soldiers were proving to a skeptical white public that they were willing to lay down their lives for the Stars and Stripes. In late May African-American regiments participated in an unsuccessful assault on Port Hudson, Louisiana, the only other position on the Mississippi River besides Vicksburg still controlled by southern forces. Although the attack failed, it changed many white northerners' attitudes toward black troops, who could no longer be written off as second-rate soldiers fit only for garrison duty and manual labor. This newfound admiration became even more pronounced in June, when black regiments helped repel a Confederate assault on Milliken's Bend designed to sever Grant's supply line (unbeknownst to the Rebels,

Grant had already abandoned the post as a supply hub). Recruited primarily to free up white soldiers for combat, these regiments were under orders "to protect the plantations that were rapidly being leased along the west bank of the Mississippi." They also may have been sent to Milliken's Bend under the mistaken assumption by federal officials that blacks were immune to southern diseases.

By the time the Vicksburg campaign got under way, Union colonel S. B. Holabird had already convinced Major General Nathaniel Banks, Butler's replacement in New Orleans, that blacks were better able to withstand the diseases that plagued the Union forts surrounding the city. Secretary of War Edwin Stanton agreed and in January 1863 issued orders for his officers in New Orleans to raise black Louisiana regiments. Two months later Halleck wrote to Grant encouraging him to use black troops to "hold points on the Mississippi during the sickly season" if Banks's experiment proved successful. The general-in-chief thought that such a scheme would "afford much relief to our armies." Surgeon General William A. Hammond agreed. Not long afterward he forwarded Stanton a report from Banks's department which seemed to prove beyond a doubt that black troops stationed in the Deep South were less vulnerable to malaria than their white comrades. The report was based on a study that had been conducted at Forts Jackson and St. Philip in Louisiana involving the Thirteenth Maine Infantry and a contingent of black Louisiana recruits. Nearly 11 percent of the soldiers from Maine were diagnosed with "intermittent or remittent fever" compared to less than 1 percent of the Louisianans, who likely enjoyed a limited degree of immunity to the local plasmodium parasites. The editors of the *American Medical Times* interpreted the results as proof that "the argument in favor of the employment of colored troops at the South, if based on their comparative immunity from the diseases peculiar to that region, is conclusive." For the surgeon general the study confirmed the "well-ascertained fact" that the African was "less liable to affections of malarious origin than the European." Hammond also believed that blacks possessed unique biological features that allowed them to "work in the rice and cotton fields of the Southern States with impunity" and avoid yellow fever infections. Blacks themselves bought into these ideas. Former slaves living on South Carolina's Sea Islands warned northern philanthropists about maladies that afflicted only whites, while black journalists pressed Washington to make greater use of "colored" troops who could resist these diseases.[14]

Northern surgeons treating African-American patients, however, discovered that most of these claims were exaggerated. While traveling with the Seventeenth Regiment Corps d'Afrique from Baton Rouge to Port Hudson, surgeon John Fish observed black soldiers suffering from malaria as well as diarrhea, scurvy, typhoid fever, rheumatism, and pneumonia. The experience challenged Fish's preconceived notions about African-American bodies. "I had supposed the black man to be peculiarly exempt from diseases due to malarial influences," he wrote, "but I should not expect to have encountered a greater number of cases of intermittent fever in a body of white troops equaling ours in number than we have actually had." Another northern surgeon experienced a similar epiphany after examining the data on 7,949 black refugees. He found that 2,776 of them, or 35 percent, had been diagnosed with either remittent or intermittent fever, which was enough to convince him that scientific claims about black immunity to malarial fevers ("so often reiterated in our text-books") were patently false.[15]

But these discoveries came too late to convince Union commanders that black and white soldiers should be treated equally, and most African-American regiments sat on the sidelines during the Vicksburg campaign. White units bore the brunt of the fighting and endured the South's final desperate attacks while waiting for the river fortress to fall. By July 1, with his men wasting away from disease and malnutrition, Confederate lieutenant general John C. Pemberton had a decision to make; he could either surrender both the army and the city and hope for decent terms from "Unconditional Surrender" Grant, or he could attempt to shoot his way through the Union lines and live to fight another day. When Pemberton asked his division commanders to find out whether or not their troops were fit enough for the second option, Brigadier General Seth M. Barton reported that only half of his men were healthy enough to fight. "The command suffers greatly from intermittent fever, and is generally debilitated from the long exposure and inaction of the trenches," he wrote. "Of those now reported for duty, fully one-half are undergoing treatment. These I think are unfit for the field."[16]

Although Barton was the only brigade commander to name malaria specifically as a reason his force was not at full strength, other Confederate commanders were almost certainly in a similar predicament. Brigadier General Alfred Cumming blamed inadequate food and "the summer sun of a debilitating climate" for the poor health of his brigade. Colonel A. W. Reynolds's brigade was in poor health because of "constant exposure" to the elements, reduced rations, and tainted drinking water. Given the high rate of malaria among Pemberton's army as a whole, it is reasonable to assume that many of the men in these brigades were infected with plasmodium parasites. In June 1863, 18.9 percent of the Confederate army at Vicksburg was officially diagnosed with malaria and 34.2 percent with "nonspecific diarrhea." These two diseases together represented nearly half of all the cases of sickness and injury. Scattered cases of smallpox, measles, and various other infections only added to the misery of the southern army. When Pemberton learned the condition of his troops, he realized escape was impossible and negotiated a surrender that marked the beginning of the end for the Confederacy. In later years Pemberton blamed the loss of Vicksburg on his enemy's superior numbers and the fact that his troops were "worn down with fatigue" after forty-seven days in the trenches. Disease, then, primarily malaria and diarrhea, was a major factor in Pemberton's defeat. Throughout the campaign malaria afflicted the Confederate army at nearly twice the rate of the Union force, mainly because of supply differences. The Army of the Tennessee had more food, clothing, and medicine and thus was better able to resist the pernicious "miasmas" of Mississippi which had wrecked the health of Williams's troops a year earlier.[17]

Malaria became a bigger problem for the Army of the Tennessee after the fall of Vicksburg. In July the rate of infection shot up six percentage points over the previous month and peaked at 16.6 percent in August. The number would have likely climbed higher still had a number of units not been moved to more active theaters, most of which happened to be healthier. The troops who were left behind suffered terribly, especially those camped along the Big Black River, where malaria soon became the most common complaint. Union commanders who believed Mississippi's "hot and sickly season" was over by late August quickly realized their mistake. The Eleventh Missouri Infantry was on the Big Black in September and suffered from intermittent fever more than any other disease. The regiment's surgeons treated patients with a cocktail of quinine and "Fowler's Solution" (arsenic), but when supplies ran low, they resorted to cinchona bark. A surgeon with the Eighth Wisconsin Infantry, also bivouacked near the Big Black, found it difficult to procure sufficient medicines for his unit. Army quartermasters only partially filled his requisitions for quinine, and as a result, the men under his care were afflicted with both malaria and diarrhea. The soldiers of the Thirty-fifth Iowa Infantry also sweated and shook with malarial fevers under the scorching southern sun. When several men in the unit contracted congestive fever, they received poor treatment as the Thirty-fifth was in desperate need of a "good Physician."[18]

Farther south, at the same time, mosquitoes were busy killing northern seamen. A yellow fever epidemic struck the Union fleet anchored at New Orleans and came close to spreading into the city itself, a nightmare scenario that could have easily destroyed the unacclimated Federal force stationed there. John H. Clarke was an assistant surgeon working at a hotel–turned–naval hospital on New Levee Street at the time. He and his colleagues received most of the one hundred cases of yellow fever and performed autopsies to confirm its presence. Clarke noticed that the epidemic was "confined to the vessels in the river" but was "spreading among them, as new ones come in from sea." Union commanders recognized the seriousness of the situation and tried desperately to keep their men from finding out about it. The commander of the West Gulf Blockading Squadron, Henry H. Bell, initially refused to quarantine any infected ships out of fear that to do so "would be utterly demoralizing to the whole fleet," but he was forced to relent in the case of the vessels that were hit especially hard. The USS *Pensacola* had forty-one cases of fever on board and ten deaths at the end of October. Interestingly, the ship's surgeon believed the disease was coming from inside the vessel and noticed that a large number of those who fell ill had worked in the engine room. The humid air, pools of freshwater, and concentration of sweaty bodies below decks created a perfect atmosphere for *Aedes aegypti* mosquitoes.[19]

Malaria afflicted the fleet at the same time. When Bell decided to err on the side of caution and send even suspected cases of yellow fever straight to the naval hospital, Clarke saw a number of patients suffering from intermittent and remittent fever. William Law, an engineer with the USS *Pinola*, was diagnosed with congestive fever (most likely *Plasmodium falciparum*) and was suffering with chills he had contracted after a visit ashore on September 22, 1863, when he had become "so prostrated that he could not walk to the ship." His physician blamed the "southern and pestilential climate." John Johnson was a seaman from Salem, Massachusetts, serving aboard the *Estrella*. Johnson had suffered from recurring intermittent fever for almost two months before "rigors," extreme thirst, and a pounding headache finally forced him to seek medical attention. John C. Clark was a blue-eyed blacksmith from Boston also suffering from intermittent fever, for which he sought treatment in September. He first contracted the disease a year earlier and had been taking quinine and arsenic with mixed results ever since.[20]

These malaria cases were overshadowed, however, by the clear and present danger of yellow fever. Few other diseases engendered the same degree of visceral fear among nineteenth-century Americans. A flurry of communiqués flew back and forth between U.S. naval officers during the summer and autumn of 1863 as the disease spread from New Orleans to Pensacola, which served as a coaling station and headquarters for the West Gulf Blockading Squadron. In August Dr. B. F. Gibbs discovered the disease on board a navy store ship anchored at Pensacola named the *Relief*. It quickly spread to a merchant ship and several "bomb-flotilla" vessels before surfacing in the nearby communities of Woolsey and Warrington. The exact number of civilians living in these villages who fell ill is unknown. Gibbs contended that the areas were mostly deserted and that only a few persons got sick. Colonel William Holbrook, who was stationed in Pensacola with the Seventh Vermont Infantry at the time, claimed that only "a few sporadic cases" cropped up in the towns in early September, but he also reported that the epidemic "raged for over two months" in the area and caused an "uncommonly large" mortality rate. However many deaths occurred, the threat of "yellow jack" hung over the heads of Pensacola's citizens in the fall of 1863 and added pestilence to the list of trials and tribulations they were forced to cope with during the war.[21]

Holbrook posted sentries around his camp and halted all "intercourse and communication" between his troops and the "sailors, marines," and "citizens" inhabiting areas thought to be infected. Mail and supply deliveries to Pensacola ceased. When a few cases appeared even on board a ship anchored near Mobile, Admiral Bell revealed his anxiety in a message sent to Captain John Gillis: "The yellow fever is in the *Colorado*, but is being subdued, I believe. God grant us this mercy." The commander found relief only when ice appeared in early December and halted the epidemic.

The chief importance of this outbreak was the damage it inflicted on the Union blockade. Ships that were needed to apprehend Confederate runners were instead quarantined or diverted to northern locales with the hope that cooler weather would destroy whatever it was in their hulls that was causing yellow fever. The USS *Albatross* steamed out of New Orleans in September to serve with the blockading fleet but was ordered back to shore two weeks later after a number of crewmen fell ill with yellow fever. A sailor on board witnessed the death of seven of his shipmates from the disease and observed "half of the Ships company" writhing in agony in their hammocks. While in quarantine at Pensacola, he also noticed the large number of other U.S. vessels alongside the *Albatross*. "All the mortar fleet that was in the river are laying here," he wrote to friend and former mate Louis Boyd. "They have all lost the best part of there crews with the fever." During the following winter and spring the *John Griffith*, the *O. H. Lee*, the *Sarah Bruen*, the *Sea Foam*, and a number of other vessels that had been infected during the outbreak were ordered North to prevent a future epidemic.[22]

New Orleans natives had been predicting that yellow jack would save their city since David Farragut's fleet had captured it in April 1862. In the weeks following the surrender of Vicksburg and Port Hudson, the last two Confederate strongholds on the Mississippi River, their predictions nearly came true. But the North's luck held. Yellow fever proved to be only a temporary setback for the navy's West Gulf Blockading Squadron. And nothing could diminish the importance of the North's victories in Louisiana and Mississippi. These triumphs were made possible by Grant and his subordinates as well as unsung heroes in Washington who were serving their country by slicing through red tape rather than Rebel regiments. No man proved more adept at navigating the federal bureaucracy than Surgeon General William Hammond. Hammond had secured his position a year earlier, thanks to the political influence of the U.S. Sanitary Commission, which lobbied the Lincoln administration to appoint a younger man to run the army's Medical Department, one who would enact sweeping changes and embrace the latest developments in medicine and disease prevention.

Hammond lived up to the commission's expectations. His reforms were one reason the North's early "improvised war" became a more organized one by 1863. On the day Farragut captured New Orleans, he inherited a moribund department that had failed to spend even the paltry sum ($2.4 million) Congress had allocated for it during the first year of the conflict. Before being ousted by his political enemies for questioning the efficacy of mercurial treatments, Hammond nearly quadrupled the department's spending and cleared many of the bottlenecks that were keeping medical supplies out of the hands of his frontline officers. Unfortunately for the Army of the Potomac, he did not receive his appointment early enough to reduce the high rate of sickness among the troops participating in the Peninsular Campaign in any significant way. But by the time the Army of the Tennessee pushed past Vicksburg and landed in Mississippi, Hammond had solved the Medical Department's worst logistical problems and issued orders for Grant's soldiers to receive regular doses of quinine.[23]

Richmond could not keep its armies equally well provisioned. By 1863 the Union blockade and occupation of broad swaths of Confederate territory were disrupting the South's supply channels. Pemberton's men were suffering from acute shortages of food, clothing, and medicine in the weeks before they surrendered—a scenario that would become increasingly familiar to both southern soldiers and civilians over the next two years. But even after its humiliating losses at Vicksburg and Gettysburg, the South was determined to fight on, which meant men and mosquitoes could continue the killing.

{6}

"The Roughest Times Any Set of Soldiers Ever Encountered"

I N THE IMMEDIATE aftermath of the epic battles in Mississippi and Pennsylvania, Union military progress slowed to a crawl. Union general William Rosecrans successfully maneuvered Braxton Bragg's Confederate army out of Tennessee before receiving a drubbing at the Battle of Chickamauga. In the East, Major General George Gordon Meade's army moved at a snail's pace after winning at Gettysburg. Lincoln believed that Meade had snatched defeat from the jaws of a total Union victory by allowing Lee's army to escape back into Virginia, but the commander in chief was in no position to dismiss one of his few generals who had proven himself a winner. Meade maintained command over the Army of the Potomac, while the president and Halleck turned their attention to the western theater. Halleck thought the states in this region, especially those located west of the Mississippi River, were ripe for the picking and could be more easily conquered (and subsequently reconstructed) than the Rebel fiefdoms back East. Lincoln agreed, and when word came that a sizable Confederate force in Little Rock under the command of Sterling "Old Pap" Price was threatening to invade Missouri, Arkansas quickly became one of the North's primary military targets.

Union planners realized that they needed to gain control over the Arkansas River in order to protect the northern part of the state and Missouri from Rebel attacks. With this goal in mind and in order to bleed the South's beleaguered armies even further, Grant ordered

Major General Frederick Steele to march twelve thousand bluecoats across Arkansas to destroy Price's army and capture Little Rock. A second Union force under James Blunt moved eastward from Indian Territory at the same time. Steele created his Army of Arkansas from regiments that had been stationed at Helena all summer—a city that one Union soldier remembered as a "malaria-stricken, disease-fostering hole"—and units transferred from Mississippi after the fall of Vicksburg. Large numbers of men in both groups were already infected with plasmodium parasites when they rendezvoused in Helena at the end of July. Two brigades from Nathan Kimball's division (Sixteenth Corps) arrived from Snyder's Bluff "suffering severely from the malarious influences of the Yazoo country." Their comrades in the Third Minnesota had contracted malaria in the same vicinity while felling trees, digging rifle pits, and performing picket duty. A number of other units sent to Steele had spent time along the Big Black River, an area the U.S. Sanitary Commission had found to be rife with disease. The parasites these men contracted in Mississippi were introduced into an environment that was already plagued with endemic malaria. The countryside surrounding Helena was dotted with "swamps" and "bayous" that were "covered with green scum in dry weather," conditions that allowed mosquitoes to thrive. In fact, malaria was so common in Helena that one soldier thought the city's residents subsisted on "corn whisky and quinine" and that their "chief occupation" was "damning the Union every day and shaking with ague every other day."[1]

The anopheles population of Arkansas quickly spread these various strains of malaria between the men in Steele's command and helped make their march to Little Rock a nightmarish ordeal. Scores of men fell out of the ranks in exhaustion, while others cast off blankets, clothing, and knapsacks to lighten the load, creating a trail of litter along the marching route. By the time Steele arrived in Clarendon, Arkansas, where he was reinforced by John W. Davidson and six thousand Union cavalrymen, malaria was epidemic in his army. Andrew Sperry was a soldier with the Thirty-third Iowa who in later years described Clarendon as the "home and head-quarters of ague in bulk and quantity." He and his fellow Iowans, who had grown up in a malarious state, believed the Arkansas variety of the disease was especially severe. One morning during their stay at Clarendon the bugler responsible for blowing the Thirty-third to sick call was shaking so severely he could not perform his duties. When two other men, including a lieutenant colonel, attempted to fill in for the ailing soldier, they too were seized with malarial shakes, leading Sperry to conclude that there "was ague in the bugle."[2]

By the time the army reached De Valls Bluff at the end of August, Steele took notice of the disease problem facing his troops in Arkansas. "The sick list is frightful, including many officers," he reported to Major General Stephen Hurlbut on the twenty-third. "More than 1,000 here present are reported unfit for duty, and about one-half of the command proper are absent." On September 1 he identified the cause of his troops' suffering: "Many of our men have been taken down with fevers, and chills and fever, lately." Leander Stillwell was a soldier with the Sixty-first Illinois who rejoined his regiment after it arrived at De Valls Bluff. He was transported by boat from Helena and along the way noticed the rugged, inhospitable conditions of the Arkansas interior: "The country along White river from its mouth to Devall's Bluff was wild, very thinly settled, and practically in a state of nature." He had been sick at Helena and arrived heavily dosed with quinine ("taken to break the fever while in the hospital"). Stillwell blamed the drug and a rainstorm for a case of rheumatism that put him out of commission for the remainder of the campaign. From the vantage point of an ambulance, however, he was in a unique position to observe the suffering of his fellow sick soldiers. "At night, the sick of each division (of whom there were hundreds) would bivouac by the side of some lagoon, or small water course," he wrote. "The attendants would prepare us some supper, and the surgeons would make their rounds, administering such medicines as the respective cases required. The prevailing type of sickness was malarial fever, for which, the sovereign specific seemed to be quinine."[3]

Malaria was rampant in Steele's army for a number of reasons. First, the troops were marching through the "swamps and marshes" of Arkansas at a time of year when mosquitoes were highly active. Nineteen thousand men and an untold number of animals clustered together presented an irresistible target for hungry female anopheles. Many of these men had spent the summer in the Big Black, Mississippi, or Yazoo river basins and were already carrying plasmodium parasites before the campaign began. Second, the Army of Arkansas was poorly provisioned. Many of the soldiers had not been issued tents and were forced to sleep in the open, making it easier for mosquitoes to feed on them. Quinine was administered irregularly because of army supply problems and the unusually high demand for the drug in the department. Steele set up temporary hospitals at Brownsville, Clarendon, De Valls Bluff, and eventually Little Rock, a move that stretched his supply of medicines to the breaking point. Requests for additional drug supplies had to be sent to Memphis, which created "long and vexatious delays." Finally, Steele's medical staff was in disarray during much of the campaign. There was a shortage of doctors—several regimental medical officers were inexplicably absent during the campaign—and at least some of those on duty were "manifestly unfit for their place." Shoddy treatment of the sick prompted Steele to take action on August 24 while the army was camped at De Valls Bluff. He ordered his regimental commanders to find out who was to blame for the "gross neglect" of ill troops. Charges of "incompetency and neglect of duty" were brought against J. C. Whitehill, assistant medical director of the army, by a group of regimental surgeons, but Steele, pressed by Confederate cavalry raids and worried about Price's next move, dropped the matter.[4]

Dysentery and the suffocating summertime humidity of Arkansas wreaked further havoc on the health of Steele's men. Lacking a steady supply of freshwater, thirsty soldiers dipped their canteens in stagnant bogs and puddles where hogs had wallowed. By the first week of September the army was at Brownsville, where it again was "encumbered with a large number of sick—near 700." Steele was forced to detach two brigades of cavalry to guard these ailing troops and keep his supply lines open. Despite these setbacks, the general drove the remainder of his army across the Arkansas River and into Little Rock. Outgunned, outflanked, and anxious to avoid Pemberton's fate, Price hastily withdrew his forces sixty miles southwest to Arkadelphia. Steele's exhausted cavalrymen only gave brief chase before retiring. In the end only 136 men in the Army of Arkansas were killed or wounded during the series of skirmishes which constituted the campaign for Little Rock, while at least 1,000 were incapacitated by disease. These results are consistent with the Union army's experience in Arkansas as a whole during the war. Thousands of northern men were poured into a state that was plagued with malaria and played only a small role in the success of the Union war effort. Far more soldiers stationed in Arkansas were prostrated by anopheles mosquitoes than were killed or wounded in battle. Between July 1863 and June 1865 over 70,000

Union troops were diagnosed with either intermittent or remittent fever, while hundreds of others contracted typho-malarial fever. Among the various Union departments that existed during the war, Arkansas was the most malarious.[5]

The prevalence of mosquito-borne disease may help explain why Steele's troops did almost nothing to impede the Confederate invasion of Missouri which went ahead in the autumn of 1864. Led by Price, this raid was designed to draw Grant's and Sherman's men away from Petersburg and Atlanta and encourage Missourians to vote the Democratic ticket in the November presidential election. At both the beginning and end of the unsuccessful campaign, the lion's share of Price's forces crossed the Arkansas River unmolested by Steele's Federals, a failure that caused "great dissatisfaction" in Washington City. Steele thought the small size of his force and the vast size of the surrounding territory made it impossible for him to bag Old Pap. A report on the department's troop strength issued at the end of October shows that only 17, 618 out of the 44,506 soldiers the general had on paper were actually available to fight. Most of the others were ill, chiefly with malaria and diarrhea.[6]

Few regiments serving in the region suffered more from malaria than the Third Minnesota. After Steele's army was routed in the early spring of 1864 while taking part in Nathaniel Banks's ill-fated Red River Campaign—a disastrous expedition but one that was not seriously affected by mosquito-borne illness—the Third was sent to garrison Pine Bluff, forty miles south of Little Rock.[7] All summer long Minnesota boys accustomed to the comfortable temperatures of the Old Northwest labored in the sweltering southern heat to erect fortifications in case of a Confederate attack. The regiment's surgeon, Dr. A. C. Wedge, was soon faced with a "violent epidemic of malarial fever," which he blamed on a "south wind" that brought in "miasma" from a nearby bayou. Wedge's analysis was probably not far from the truth; in all likelihood the breeze was extending the flight range of an untold number of female anopheles hatched in the bayou which were searching for a blood meal. Many of the men in the Third had spent the previous summer in the Yazoo River Valley with Grant's army and participated in Steele's march from Helena, so they were probably already infected with plasmodium parasites by the time they got to Pine Bluff. A contingent of "several hundred" fresh recruits arrived from the North in April, and soon 80 percent were severely ill with "malarial fever." Wedge was without quinine, so the unit continued to suffer until Minnesota authorities intervened (after a series of letters were sent to the state's governor and senator) and sent down a large supply of the drug. The Third likely suffered from gastrointestinal ailments at the same time given that the local water supply was deemed filthy. When several companies were furloughed in mid-August and arrived in De Valls Bluff on the way home, Brigadier General Christopher Andrews described the scene as "touching." "Most of them looked pale and thin," he recalled. When the regiment continued to waste away in the Arkansas swamps for another eight months, Minnesota's governor, Stephen Miller, penned a strongly worded letter to Major General Edward Canby on their behalf, requesting a transfer. "For more than eighteen months the regiment has been engaged upon fatigue duty on fortifications and railroads," he complained. "During that time 150 of their number have died of the fever incident to that terrible climate, and the survivors and their executive now ask that they may have an opportunity to die at the post of honor." The men of the Third never got another chance to die honorably. Five days after Miller wrote his letter, Lee surrendered at Appomattox.[8]

Other regiments also wasted away in Arkansas, especially those stationed in Helena. The Sixth Minnesota was made up of more than 900 healthy soldiers when it left Cairo, Illinois, for Helena in June 1864, but by September only 144 men were fit enough to fight. The regiment was camped on the banks of the Mississippi River and surrounded by muck-covered marshes that served as breeding pools for mosquitoes. Two weeks after the Sixth arrived, the hospitals at Helena were overflowing with patients. The unit's surgeon, Dr. W. P. Belden, described many of the men he treated as "jaundiced and sallow" and recommended that the camp be moved to a healthier location. Hundreds of men who never fired a shot on a battlefield were evacuated to Union hospitals in Memphis and St. Louis. Dysentery and malaria were rampant and continued to erode the health of the regiment until it was granted a furlough at the end of September. Charles Johnson was a soldier with the Sixth. Twenty-five years after the war ended, he was still angry with the army for its decision to send his unit to Helena. Johnson thought it a "monstrous outrage" to send any "unacclimated Northern regiment" to such an unhealthy post. The appalling numbers of sick reveal the source of Johnson's lingering resentment. Every month from July to October 1864 at least 400 of his comrades were ill, and during September—the time of year when mosquitoes were most active in many regions of the South—the number peaked at 654. Alfred J. Hill was a soldier with Company E who watched disease turn the Sixth into a "shadow of its former self." He remembered the pitiful look of "the companies as they marched to dress parade," many with only "half a dozen men in line." Malaria and other maladies inflicted more damage on the regiment than Rebel minié balls and limited its combat readiness. When the infamous Confederate raider Joseph Shelby and a group of southern cavalrymen began murdering black soldiers and plundering plantations near Helena in August, the Sixth and another Union unit from Missouri could only find 400 men healthy enough to respond.[9]

Black troops stationed in the area also suffered from disease. The Fifty-sixth U.S. Colored Infantry was stationed in Helena for over a year, with many of its companies camped close to the brushy ravines and stagnant pools that surrounded the city. A physician sent by the U.S. Sanitary Commission to inspect the area found that most of the seventy patients in the general hospital were suffering from malaria. The men of Company D pitched their tents near a marsh and were devastated by remittent fever; forty-six of the seventy-nine soldiers in the company fell ill. Bacteria-infested water, unsanitary conditions, and mosquitoes continued to make life miserable for the Fifty-sixth even after Union commanders recognized the extraordinary sickliness of the post. In January 1865 Halleck advised the commander of the Department of Arkansas, Major General Joseph J. Reynolds, to abandon Helena "both on account of its unimportance and its unhealthiness." Unfortunately for the troops in the area, Halleck's advice was either overlooked or disregarded, and the post remained occupied for the duration of the war.[10]

Although records for the Confederate armies that operated in Arkansas are incomplete, the available evidence suggests that malaria was also a significant problem for the southern troops assigned there. Early in the war John Walker's Texans spent time near Clarendon and were hit hard by "fever and ague." Over 50 percent of the division fell ill, and "several regiments" could not even muster enough men for guard duty. Junius Bragg was a surgeon for a Confederate unit stationed along the Arkansas River. "This Ark. Post is a *vile* place," he wrote to his future wife, Josephine. "It is so unhealthy here in the summer season, that nothing can live except mosquitoes. I am credibly informed by the oldest 'inhabitant' that the *snakes* have *chills* here in the summer." In 1863 Bragg was sent to Pine Bluff, where nearly "every one" under his care contracted "chills." The doctor himself also fell ill with malaria. Nor was he the only southern physician in Arkansas to be infected with the disease. Dr. William McPheeters was with Theophilus Holmes's force when it launched an unsuccessful attack against the Union garrisons at Helena—a campaign intended to draw northern troops away from Pemberton's forces at Vicksburg—and developed intermittent fever on the retreat back to Little Rock. His ailment periodically returned and at one point forced him to spend an entire day in bed.

The winter after the campaign McPheeters and his fellow surgeons created a military medical association for the purpose of exchanging ideas and information. During their first meeting three papers were presented "on the peculiarities of intermittent fever." Returns from the Confederate District of Arkansas after the Battle of Helena suggest a high rate of sickness. The three divisions that constituted the district had almost 32,000 troops on paper, but less than half that number were actually available to fight. The returns from the districts of western Louisiana and Texas reveal a similar divide between paper strength and actual numbers of active-duty troops.[11] With three times fewer men eligible for military service than the North, the South could ill afford to lose so many soldiers to disease. But for both Confederate and Union forces operations in the trans-Mississippi region during the summer and fall months were risky, in no small part because of mosquito-borne diseases.

As malaria steadily eroded the health of Steele's men in Arkansas, Confederates in Texas were dying of yellow fever. During the winter of 1862–63 John B. Magruder had successfully driven off the Massachusetts regiment that had been sent to capitalize on Renshaw's victory and occupy Galveston once the sickly season was over. Having successfully recaptured the Confederate port city, Magruder proceeded to fortify it with more than three dozen cannon placed in various forts erected on Galveston and Pelican islands and at Virginia Point. Before his troops were decimated by the disease, the Virginia-born general also considered yellow fever to be one of Texas's military assets. In May 1863 he transferred two thousand troops "from the Rio Grande frontier" to Louisiana, confident that his western flank was safe from attack during the "yellow-fever season."[12]

The following year, however, Magruder learned that the scourge of the South was one of its least dependable allies. By the summer of 1864 conditions inside the city were deplorable. Confederate soldiers were bedeviled by food and supply shortages and were owed back pay. Civilians suffered as well. One Confederate soldier noted in his diary that destitute local children often hung around his camp, begging for something to eat. In August rumors circulated that yellow fever was present in the city, prompting Magruder to quarantine ships arriving from areas known to have problems with the disease, such as the West Indies and Mexico. The scene got even uglier when several hundred frustrated and nervous soldiers mutinied and surrounded a house where Magruder and his officers were being entertained by the ladies of Galveston. The general was able to talk his way out of the situation but was probably relieved when he was transferred later in the month; by the middle of September Galveston was in the throes of a full-blown yellow fever epidemic. The first official death attributed to the disease was that of a forty-year-old minister from Germany who perished on September 5. Soon other men, women, and children, including a substantial number of German immigrants, also died from the disease. The outbreak "caused considerable excitement among the troops," and "a number of them" tried to desert their posts under cover of darkness on September 16, the night before a lockdown of the island went into effect. Brigadier General James Morrison Hawes, the officer in charge, asked the residents of Houston to send nurses to attend to the sick and dying, but because of an epidemic in their own city, they were unavailable to come.

By the end of September fifteen cases and three deaths were reported in Houston. A few weeks later more than one hundred people were ill with the disease. To escape the epidemic Major General John Walker—who had replaced Magruder in August as commander of the District of Texas, New Mexico, and Arizona—hastily moved his headquarters out of the city. But despite the large number of sick people in Houston, few deaths occurred. The local newspaper reported that the disease was "of the mildest type ever known in that city" and that few perished who were able to receive prompt medical attention. The epidemic, however, lasted longer than most. In late November warnings were still being issued for unacclimated people to avoid the city, and a month later the Confederate States District Court was forced to suspend its proceedings because of the disease.

The yellow fever outbreak that occurred in Galveston at the same time was much more serious. The disease claimed the lives of the city's most prominent citizens and its most humble. The Reverend J. M. Goshorn, rector of Trinity Episcopal Church, and his wife died the same month that Bend, a thirty-year-old slave, was buried. Inside the city law and order rapidly deteriorated, and thugs took advantage of the confusion to prey upon helpless civilians. Cornelius O'Connor's house was burglarized while he was away, and the woman he had left in charge of his property was murdered during the robbery. Other murders occurred as well. When an Englishman named Baker shot and killed a Confederate soldier for plundering his chicken coop, the other members of the soldier's company took revenge by stabbing Baker and hanging his corpse on a post in front of his house. Confederate commanders were eventually able to regain control by instituting regular patrols, but yellow fever, rival attacks by Federal and Rebel forces, lawlessness, and supply shortages made life under the Confederate government worse for most citizens in Galveston than it had been under the old Constitution.

The chaos afforded the Union fleet patrolling the waters off Galveston an opportunity to recapture the city, but even the toughest salts were not about to expose their crews to the dreaded yellow jack. Rear Admiral David Farragut, in an apparent retreat from his earlier policies, issued orders to his commanders off Galveston to steer clear of the city and to avoid picking up passengers.[13]

By the time the first frost had halted the Galveston epidemic in late November, 111 soldiers and 158 civilians were dead. Nearly 20 percent of the casualties in the army occurred among German immigrants who had grown up in areas where yellow fever was nonexistent and as a result had not developed an immunity to it. These were men such as twenty-five-year-old John Eickelberg, who lived in Austin County prior to the war and served as a musician in Elmore's regiment before a mosquito bite ended his life. Or Private Zitzleman, who died as a member of Wilke's Battery and was buried in the potter's field in Galveston. Confederate soldiers from England, France, and Ireland also perished during the outbreak, along with dozens of native Texans. Walker believed the outbreak had been a "great drawback" to his defense strategy for Galveston. His overall plan for Texas depended on occupying key locations and using interior lines to transfer troops to threatened points, but he was forced to pull soldiers from Sabine Pass to prop up the weakened command at Galveston. Fortunately for the

Confederates, by the autumn of 1864 Union military planners were uninterested in committing large numbers of men to an area with little strategic value which was periodically beset by dangerous plagues. The attack Walker had prepared for never materialized, but the shock of the 1864 Galveston epidemic lingered after the war was over. Union forces occupying the city during the summer of 1865 instituted a quarantine and strict sanitation measures in much the same way Butler had three years earlier in New Orleans. The next major outbreak in Galveston would occur two years later and kill more than 1,100 people.[14]

Meanwhile, across the Gulf of Mexico yellow jack again hit the Union fleet anchored at New Orleans. This time nearly 200 sailors on 25 different vessels fell ill, resulting in 57 deaths. The entire crew of the USS *Tennessee,* a captured Confederate ironclad, was evacuated, while repairs on other damaged vessels came to a virtual standstill. The surgeon attached to the USS *Ossipee* ordered the destruction of all clothing, blankets, and mattresses belonging to a number of crewmen who had come on board from another infected vessel, the USS *Union.* When yellow fever again surfaced in nearby Pensacola, it was blamed on a steamer that had stopped for coal in Key West, which was experiencing its own outbreak at the time. A handful of workers at the naval hospital in New Orleans also got sick (three died), along with five civilians, but, fortunately for the thousands of Union soldiers stationed nearby, the disease did not spread through the city and died out when cooler weather arrived.[15]

While northern sailors were dying in New Orleans in 1864, dengue fever and yellow fever struck Key West and the Dry Tortugas Islands. According to one observer, at least "five hundred" people at Fort Jefferson on the Dry Tortugas were afflicted with dengue fever, which caused a pall of gloom to fall over the community. The sweltering temperatures of southern Florida turned the fort into a giant brick oven that broiled patients who were already burning with fever. In Key West yellow fever was "raging" with such ferocity that supposedly "even the old acclimated residents" fell ill.[16]

Thirty-three soldiers stationed at Key West died, including Brigadier General Daniel P. Woodbury, who was "buried with military and Masonic honors" on August 17, 1864. The fever hit the Second U.S. Colored Troops particularly hard and sickened nearly all of the outfit's white officers. John Augustus Wilder was one of those officers stationed at Fort Taylor. He observed that at one point during the epidemic "30 or 40 new cases" appeared each day "among the sailors & others." Wilder was appalled by the amount of suffering he witnessed. "It seems hard to realize that so delightful & sunny a land on which the smile of God seems visibly to rest should be so deadly & deceitful," he wrote to his brother-in-law. "Is this indeed Paradise or is it Hell?" For many black soldiers stationed at Key West, southern Florida seemed more like the nether regions. Isaiah Johnson was an African-American farmer and musician from Norfolk, Virginia, who perished at the end of July. His comrade, nineteen-year-old George Washington, had grown up in Mathews County, Virginia, and worked as a waiter before he died on a hospital bed the same month. William Stripling and John Baker were both from South Carolina and joined the Second after it was transferred to Key West to fight for their relatives' freedom. The two men were early victims of the epidemic and died in May. Basil Berry enlisted in Washington, D.C., as a substitute for James Bradley and died at a hospital in Key West. In the end 7 other black troops in the Second died, and 132 men got sick. The regiment also suffered from smallpox, typhoid fever, chronic diarrhea, dropsy, and consumption while stationed in Florida.[17]

The navy ordered its vessels to avoid Key West as it had during the 1862 epidemic. Those ships unlucky enough to be moored in the area as the pestilence spread soon became infected, prompting Rear Admiral Theodorus Bailey to order a general evacuation. "My worst fears have been more than realized," he wrote in July, "and for more than two months the disease has held its course without abatement and is now as virulent as at any time." The crews of dozens of ships under his command were laid low. Every sailor aboard the *Nita* got sick except for two officers, and at least ten men died. The USS *San Jacinto*, which had been the scene of so much suffering during the 1862 epidemic, was again ordered north, this time to Portsmouth, New Hampshire. The *De Soto*, the *Huntsville*, and the *Merrimac* (not to be confused with the ship of the same name that became the CSS *Virginia*) were also ordered to northern ports as men on board the *Honduras*, the *Honeysuckle*, the *Iuka*, and the *Marigold* fell ill and died. Other ships were sent to Tampa, but yellow fever followed. Bailey himself was eventually relieved of his command by Gideon Welles, who recommended his subordinate "return north for the purpose of recruiting your health." During the summer of 1864 Florida's *Aedes aegypti* mosquitoes accomplished what would have been impossible for the anemic Confederate navy at the time, scattering the Union's East Gulf Squadron like sticks in the wind. Fortunately for Washington, the South was too weak to take advantage of the chaos, and by October the fever had burned itself out.[18]

More important, the outcome of the war was being decided at the time in Georgia and Virginia rather than Florida. Mosquitoes swarmed soldiers serving in these states too. Sherman's army suffered with malaria in the summer of 1864 as it maneuvered around Joe Johnston's Confederates toward the important southern supply hub of Atlanta. In July and August Cump's surgeons diagnosed over twelve thousand cases of the disease, but thanks to the Union army's ample supplies of medicine, few of these cases proved fatal. Surgeon C. S. Frink watched malaria sicken several Union regiments that had bivouacked "in a strip of heavy timber" rising out of the Georgia clay. Although "free doses of quinine" quickly improved the health of the men in these regiments, recrudescent fevers and new infections continued to dog them for the remainder of the campaign. Had Joe Johnston been able to cut Sherman's supply line successfully, malaria would have been a much more significant problem for the combined armies that captured Atlanta. Johnston, who would be replaced in July by the recklessly aggressive John Hood, had his own struggles with malaria. He was losing an average of thirty-four Confederates per month to the disease. Jasper James was a soldier with the Army of Tennessee during the campaign and by June 1864 had already suffered "three agues." "We have had as you know one of the roughest times any set of soldiers ever encountered in this climate," he wrote to his parents. "The consequence is, great numbers are sent of sick daily, judging from a company the army must have reduced greatly." Reinforcements that Johnston received from Mobile, Alabama, put his troops at further risk of getting sick.[19]

Mobile was one of the most malarious posts in the Confederacy. In one nineteen-month period, from January 1862 to July 1863, the Confederacy had a mean strength of 6,752 troops stationed in the city, with 13,668 cases of malaria (10,500 "intermittents" and 3,168 "remittents"). William Spotswood, the head of the Confederate navy's medical department, issued a report in November 1863 which revealed the extent of the malaria problem in Mobile. With "bilious," "remitting," and "intermitting" fevers sapping the strength of Confederate crews on ships anchored in the bay, especially those situated "amidst the marshes," Spotswood recommended that the vessels be moved to healthier locations. In August 1864, as Sherman inched closer to

Atlanta, David Farragut led a Union armada into Mobile Bay against three forts and a small southern flotilla. As he shouted the famous phrase that would thrill subsequent generations of American school-children—"Damn the torpedoes! Full speed ahead"—Farragut faced Confederate sailors who were either ill or recovering from malaria. Southern military personnel on shore were also suffering from the disease. Ross Hospital in Mobile kept records on 1,601 of its patients in 1863, and among them 445 had been admitted for intermittent fever. In July and August 1864, around the time Farragut was attacking, Ross received 706 patients, of whom 336 were suffering from "intermittent, remittent, or congestive fever." And that autumn "six-sevenths of the officers and men" in some units were prostrated by "chills and fever," prompting the major general in charge to warn Richmond that the large sick list had left the city all but defenseless. Months later, seeing the hopelessness of the situation and realizing that reinforcements were not forthcoming, the handful of sickly Confederate garrisons guarding the city of Mobile melted away when General E.R.S. Canby's Federal forces finally arrived near the end of the war. With its appalling disease rate during the sickly season, wartime Mobile was no different from Baton Rouge, Galveston, Pensacola, Vicksburg, or any number of other towns in the Deep South where mosquitoes made life dangerous for soldiers on both sides.[20]

But mosquito-borne disease posed the greatest threat to military personnel serving west of the Mississippi River. In fact, the two states where mosquitoes were the most troublesome were both within the region the Confederacy dubbed the Trans-Mississippi Department. For the Union the Department of Arkansas held the ignoble distinction of being its most malarious. Between 1863 and 1865 there were an astonishing 1,287 cases of malaria each year for every 1,000 northern soldiers assigned to the state. Texas experienced more yellow fever outbreaks than any other state in the Confederacy. Southern soldiers stationed in port towns such as Galveston, Lavaca, and Sabine City burned with fever and spewed black vomit before drawing their last breath. But west of the Mississippi malaria was a far more pervasive problem. Repeated infections of the most common form of the ail-ment, *Plasmodium vivax*, further weakened the immune systems of men already suffering from deadlier maladies such as chronic diar-rhea and typhoid fever and contributed to the region's high rate of disease-related fatalities. *Plasmodium falciparum* was by itself lethal and occurred almost exclusively below the thirty-fifth parallel, where the bulk of the territory in the trans-Mississippi theater lay.

Soldiers transferred to Arkansas from Mississippi at the conclusion of the Vicksburg campaign brought with them various strains of malaria which were quickly spread by armies of anopheles. Compounding the problem were crowds of slaves who sought the protection of Union armies and inadvertently spread plasmodium parasites in the process. Regiments sent straight from the North as reinforcements, such as the Sixth Minnesota, were quickly overwhelmed by disease. African-American soldiers, believed to be immune to the southern sickly season, also fell ill. Dying on a sickbed rather than a battlefield violated the unwritten rules of Victorian masculinity and was a fate most soldiers thought shameful. But for thousands of young men who served in the states west of the Mississippi River, this was the sum of their Civil War experience. By continuing to pour troops into the region despite its limited strategic value and high rate of sickness, the federal government reduced the size of its armies in the East—where the outcome of the war was being decided—and caused needless suffering for innumerable soldiers and sailors.[21]

Of course, the men in blue and gray who were slaughtering one another in the East were not exempt from the ravages of mosquito-borne disease. Neither were the tens of thousands of southern civilians caught in the crossfire. The Union blockade and occupation choked off their access to essential supplies, including quinine, and made escape during the sickly season a difficult ordeal. By the summer of 1864 Confederate patriotism was waning, but as events would show, a handful of southerners were still willing to sacrifice everything—including the lives of innocent men, women, and children—in order to secure their nation's independence.

{7}

Biological Warfare

O N A HUMID DAY in early August 1864, a clerk working in an auction house located a few blocks from the White House looked up and saw a diminutive man with a swarthy complexion dragging several trunks through the front door of his office. The man introduced himself as Mr. J. W. Harris and asked the clerk if he would be interested in auctioning off the trunks, which he claimed contained twelve dozen shirts. Assuming the man was just another northern sutler hoping to turn a quick profit while on his way home from the battlefields of the South, the clerk agreed and advanced the stranger one hundred dollars. When he later opened the trunks, he found more shirts inside than he had bargained for crammed in together with several coats. Initially skeptical that he could make a profit off the merchandise, he bundled the shirts into packages of a dozen and sold them the next day for $142.90.[1]

Unbeknownst to the clerk, this seemingly innocent business transaction was in fact part of a nefarious plot designed by Confederate agents to launch a biological attack against the city of Washington. The man who had identified himself as J. W. Harris was actually named Godfrey J. Hyams, and the clothing he sold had been worn by yellow fever victims before they died. The agents working with Hyams hoped the infected fabric would spread yellow jack throughout the District of Columbia and kill officials serving at the highest levels of the federal government. The science behind this attack was of course specious given that only *Aedes aegypti* mosquitoes can transmit the yellow fever virus. But many nineteenth-century physicians, unaware of the link between insects and disease, believed "fomites," or infected clothing and bedding, could spark epidemics.

Hyams's boss was a Kentucky physician and Confederate zealot named Luke Pryor Blackburn, whose work with yellow fever patients in Mississippi before the war had earned him an international reputation as an expert on the disease. Blackburn had held a variety of jobs with the Confederate government before Mississippi governor John Pettus sent him to Canada to obtain supplies for blockade-runners. While in exile, the doctor decided to disregard his Hippocratic oath and use his knowledge of the "scourge of the South" against its enemies. He found an ally in Hyams, a cobbler who had fled Helena, Arkansas, and settled in Toronto to escape the war but later decided he needed to do something to help the struggling Confederacy. The two were introduced by other southern agents headquartered in Canada, and Hyams agreed to help carry out Blackburn's plan, which involved using trunks full of contaminated clothing as biological weapons. They met in December 1863 at the Queen's Hotel in Toronto, where the doctor told Hyams to go home and await further instructions.[2]

In order for his scheme to succeed, Blackburn needed access to a yellow fever epidemic. His chance arrived in the spring of 1864 when an outbreak surfaced in Bermuda. He immediately volunteered his services to the Bermudan government, which was grateful to receive assistance from such a well-known American physician. During his stay on the island Blackburn collected the clothing of terminally ill patients, which he packed into several large steamer trunks. In July he arrived in Halifax, Nova Scotia, trunks in tow, and sent word for Hyams to meet him. Hyams helped unload eight trunks and a "valise" off a steamship and was instructed to take the five trunks containing contaminated clothing to New Bern, North Carolina, and Norfolk, Virginia, both of which were occupied at that time by Federal troops, as well as Washington, D.C. Blackburn had a special purpose in mind for the valise. He wanted it delivered to the White House under the pretext of being a gift for the president from an anonymous admirer. Inside the small suitcase were several expensive dress shirts, but it had previously been packed with clothing taken from yellow fever and smallpox victims. Unnerved at the prospect of showing his face at the front door of the executive mansion, Hyams refused to deliver the valise but agreed to convey the trunks to their intended targets. He bribed a local steamboat captain to smuggle the luggage into Boston and from there accompanied it down the East Coast, through Philadelphia and Baltimore, placing the clothing in several other trunks he purchased along the way. When he arrived in Washington, Hyams sold five of the trunks to the auction house and four others to a sutler named Myers, who was conveniently headed to both Norfolk and New Bern. Having fulfilled his part of the agreement, the Arkansas shoemaker returned to Canada to collect the fortune Blackburn had promised would be his once the mission was completed. The doctor temporarily appeased his underling with one hundred dollars and promises of future payments and headed back to Bermuda to collect more infected clothing for use against the hated Yankees.[3]

During this second trip Blackburn filled three trunks with "blankets, pillow cases, drawers," and other material "stained and soiled" with black vomit and left them on St. George's Island in the care of a hotelier named Edward Swan. The Kentucky physician was worried that a northern cold snap might render his weapon useless and wanted to wait until the spring of 1865 to launch another attack. According to witnesses who later testified at a government hearing, Blackburn instructed Swan to ship the trunks to New York City and to collect compensation for his efforts from the Confederate office in Bermuda. The doctor then took passage on a steamer bound for Canada and never returned to the islands.

In April 1865, as the Confederacy itself was falling apart, the whole plot began to unravel. Tired of trying to collect the large sum of money he had been promised by Blackburn, Hyams decided to spill his guts at the U.S. District Attorney's office in Detroit in return for a pardon. Several days later in Bermuda, a Confederate agent tipped off authorities about the trunks at Swan's boardinghouse, prompting a police raid and a full-scale investigation. As details of Blackburn's activities emerged, it became clear that his plan to kill both northern civilians and soldiers with yellow fever had been approved at the highest levels of the Confederate government. According to Hyams, the hundred dollars he received for the Washington mission had been paid by Colonel Jacob Thompson, a Confederate operative who was involved in a plot to disrupt the election of 1864 by starting fires and sowing mayhem in a number of northern cities. Thompson's secretary, a man named

William Cleary, who was also a close confidant of Jefferson Davis, admitted that he and his boss knew about Blackburn's plan but denied any direct involvement. Davis himself almost certainly knew about the plot. He and Blackburn had been friends before the war, and during the winter of 1864 the Confederate president received two letters from an Episcopal clergyman and southern agent named Kensey Stewart which contained details of Hyams's mission. God, warned Stewart, would not be pleased with southerners if they employed such wicked tactics against their fellow men. Despite these warnings, Davis did not try to stop Blackburn, who continued his quest to start an epidemic among the vulnerable northern population.[4]

The Confederate government's support of Blackburn's plot seems particularly heinous in light of the events that took place along the North Carolina coast in 1864. That fall yellow fever struck Union-occupied New Bern with a vengeance. The first officially documented case appeared on September 9 with the death of Sergeant Mason Rogers of the Fifteenth Connecticut. Fortunately, before the war the regiment's surgeon, Dr. Hubert V. C. Holcombe, had spent time among yellow fever victims in Vera Cruz, Mexico, and recognized the disease; none of his colleagues had ever seen it. When he reported his findings up the chain of command, the Union leadership, fearful of creating a panic, immediately denied that there was a problem. Surgeon D. W. Hand, the medical director for Union forces in North Carolina, was not yet ready to acknowledge that yellow fever was present and criticized Holcombe for unnecessarily causing a stir. By one account Hand had even ordered a cover-up when a case appeared earlier in the summer. But when the medical director finally accepted the truth and informed his superiors of the problem, he encountered similar hostility. General Innis N. Palmer, the commander of the department, launched into a tirade of "scorn and abuse" and accused the director of being ignorant of southern fevers. Although Palmer eventually relented and accepted that yellow fever had broken out in his command, both he and Hand tried to keep the news from the press. Word of the outbreak leaked to the enlisted men anyway, creating considerable "anxiety." Hand also did a poor job of keeping his surgeons informed about the threat they were facing. Two weeks after Rogers's death, Union sailors as far away as Beaufort knew yellow jack was present in New Bern, even though some doctors within the city still could not identify the disease. Cases began to pile up in the area hospitals, and soon a full-blown epidemic was raging through the city. Northern soldiers who had heard horror stories of the scourge of the South were terrified.[5]

Union officials enacted a strict quarantine and burned pine branches and barrels of tar and turpentine in the streets in an attempt to purify the poisonous air, leaving the houses in the city covered in black soot. Inside soldiers discovered entire families lying dead in their beds or on the floor. New Bern quickly took on the appearance of a ghost town as houses were tightly shuttered and stores closed. Buildings thought to be infested with pestilence were burned to the ground, and in a short time the only activity on the streets came from the "dead wagons" sent to collect corpses. Army hospitals overflowed with patients, and those who could bear the journey were ferried to facilities in Morehead City and Beaufort. Healthy soldiers too were hastily transferred out of town and given furloughs, although some units remained behind to implement sanitation measures, maintain order, and keep the city from falling back into enemy hands. Because blacks were believed to be largely immune to yellow fever or perhaps simply because they were considered expendable, African-American regiments were moved in to keep the bonfires burning and spread "two hundred carloads of lime" around town. Fifty-eight of their number fell ill, and 15 of them died. The disease spread to Beaufort, where a local hotel, thought to be a source of the pestilence, was shut down to keep a panicky mob from burning it. Before the first frost of November halted the activity of the local *Aedes aegypti* population, 763 Union soldiers in the area contracted yellow fever, and 303 died from it. Brave medical officers who stayed to treat their patients were hit especially hard. Of the 16 infected with the virus, half of them died.[6]

The epidemic had a significant psychological impact on Union commanders and reinforced their belief that the south Atlantic region was an unhealthy theater of operations. As yellow fever wreaked havoc on civilians and soldiers in New Bern, word leaked out that it was also present in the Confederate strongholds of Wilmington and Charleston, creating considerable consternation within the Union's Department of the South. Orders went out to district commanders to maintain strict cleanliness in camps (which included the liberal use of lime) and to quarantine Rebel deserters and refugees before allowing them behind Union lines. Above all, commanders were instructed to keep the situation quiet to avoid a panic among the enlisted men. Northern troops who had come of age reading about the terrifying yellow fever epidemics that hit the South year after year during the 1850s suddenly found themselves occupying an area where the risk of contracting the disease was all too real. These fears seemed justified when yellow jack razed New Bern and surfaced in neighboring Charleston, South Carolina. There the disease swept through facilities housing Yankee prisoners, causing fear and uncertainty among guards and captives alike. Confederate authorities moved quickly to evacuate the vulnerable prisoners but were too late in some cases. According to John V. Hadley, a northern prisoner from Indiana, at least thirty of his fellow POWS perished before they could be moved out of town. A number of southerners died as well, including Colonel D. B. Harris, General P.G.T. Beauregard's chief engineer.

Although both northern and southern newspapers reported that Charleston was in the throes of a major yellow fever epidemic, the outbreak was limited in scope and curtailed by frosts before it could get out of hand. The same season a few cases of the disease also showed up in Wilmington, North Carolina. Whether or not the 1864–65 Union attacks against Fort Fisher, which led to the capture of Wilmington, were planned for the winter months in part to avoid the yellow fever season is not known, but sources show the disease worried U.S. troops occupying the city in the summer after Appomattox. In July 1865 Grant issued orders for the occupation forces in New Bern and Wilmington to keep the cities clean and to head for the "pine woods" in the event yellow fever appeared.[7]

Union soldiers were not the only ones devastated by the New Bern epidemic. Southern civilians watched in horror as roughly seven hundred of their neighbors died in the space of a few short weeks. The most unfortunate victims were those unable to obtain treatment from overwhelmed U.S. Army surgeons and as such, were forced to suffer in solitude. When a group of concerned local citizens banded together to address this problem and provide care for the sick and dying, half their number perished. Davis and his agents almost certainly knew about the New Bern outbreak (both the northern and southern press picked up the story) and the misery it was causing for their own people. If so, they would have assumed that Blackburn's scheme had been successful and that the trunk Hyams sold to the southbound sutler was responsible for the epidemic. The people of New Bern certainly believed as much. When a postwar government investigation exposed Blackburn's part in the plot, a local newspaper editor summed up his fellow citizens' outrage: "This hideous and long studied plan to deliberately murder innocent men, women, and children, who had never wronged him [Blackburn] in any manner, is regarded here as an act of cruelty without a parallel—a crime which can only be estimated and punished in the presence of his victims in another world." In spite of the grisly reports coming out of North Carolina during the autumn of 1864, Stewart seems to be the only official who raised any objections to Blackburn's activities. By all appearances the Confederate government was willing to wage biological warfare, even at the risk of killing its own citizens, in order to secure southern independence.[8]

Unfortunately for southern civilians, the United States government was indirectly waging a biological war of its own against the Confederacy. The naval blockade, first enacted by Lincoln in April 1861, squeezed off more and more of the South's medical supplies as the war wore on. Few drugs were in higher demand than quinine, which was believed to be a cure-all for a range of ailments but was effective only against malaria. Some parts of the South experienced shortages of the medicine as early as the first summer of the war, and prices climbed each year thereafter. Wartime profiteering and the depreciation of Confederate currency contributed to the problem. In one Carolina drugstore quinine was selling for $4.00 per ounce during the summer of 1861. Two years later it fetched $22.25 per ounce in Richmond. And late in the war in some areas the drug was selling for as much $600 per ounce.[9]

As a result of these shortages and high prices, a black market developed, and innumerable methods were devised to smuggle quinine into the South. Women were especially adept at sneaking the drug through Union checkpoints, and some of the most colorful tales of the war revolved around their efforts. Containers of medicine were hidden underneath the enormous hoop skirts that were fashionable in the 1860s. Virginia Moon, or "Miss Ginnie" as she was known to friends, was just sixteen years old when her native Tennessee left the Union. Anxious to assist the Confederate war effort, Moon traveled to Ohio, where she had previously been enrolled in a private school, and returned to her home state with "a row of quinine bottles" and "a row of morphine bottles" attached to her "underskirt." A similar stunt was attempted by Louisa Buckner, the niece of Montgomery Blair, the postmaster general of the United States and a trusted member of Lincoln's cabinet. She borrowed money from her uncle, ostensibly for the purpose of buying groceries, but used it instead to purchase over one hundred ounces of quinine, which she promptly had sewn into her dress. Buckner was arrested before she could escape Washington, D.C., and spent a short time in the Old Capitol Prison before her famous relative used his political clout to obtain her release. Nor was she the only woman with connections to the White House involved in quinine smuggling. Mrs. Clement White was the half-sister of Mary Todd Lincoln and a guest at the president's first inauguration. Once the war started, however, White decided to return to Alabama and took a large amount of quinine with her. Other women smuggled the drug in more creative ways. "Mrs. N. J. Reynolds," "Miss Shuller," and "Miss Maggie Oliver" were arrested in Vicksburg when quinine was discovered under false bottoms in their luggage. And in one case the medicine was found inside the head of a child's doll. Episodes such as these helped erode the gender stereotypes of the period, as middle-class ladies with vials of quinine sewn into their dresses demonstrated skills and daring beyond those envisioned by the cult of domesticity.[10]

Men smuggled their share of quinine as well. Authorities nabbed William T. Wilson on a steamship in Baltimore and charged him with "transporting contraband letters and goods to and from Virginia." At the time of his arrest Wilson was found carrying a "quantity of quinine." A "Doctor Moore" was also arrested for illegally transporting the drug and was sentenced to hard labor on Ship Island, Mississippi, which had recently been converted into a Union prison camp. He was forced to wear a ball and chain for his crime. In nearby Louisiana, Captain Trask of the Fourteenth Maine Infantry captured a Confederate paymaster carrying "a large quantity of quinine" along with $20,000. And in Memphis the drug was stuffed inside the intestines of dead livestock in order to move it undetected past Federal troops and out of the city.[11]

But these individual efforts produced only a fraction of the quinine needed by the South. A larger quantity was brought in by blockade-runners, which supplied the Confederacy with a range of much-needed items throughout the war.[12] A glance at the cargo manifests of runners operating out of Bermuda shows large amounts of quinine being smuggled into southern ports as early as 1862. The British-built screw steamer *Minho* had seven cases of the medicine (over 1,200 ounces) when it left St. George's in September. Another runner, the *Herald*, transported thirty cases the following month, along with gunpowder, saltpeter, and other war materiel. These shipments netted enormous profits for investors. The Anglo-Confederate Trading Company, for example, which ran ships between the Bahamas and North Carolina, was able to pay its original shareholders a 2,500 percent return on their investment in the autumn of 1864. A list of items smuggled into Wilmington on board an Anglo-Confederate vessel shows that quinine accounted for some of the firm's success. Northerners also made money off the drug. Union blockade ships intercepted millions of dollars worth of contraband goods, including medical supplies, which were sold off at auction houses in northern cities. Sixteen hundred ounces of "Pelletier's Quinine" and a few hundred ounces of opium seized from a Rebel ship fetched close to $7,500 for one shrewd seller. Another batch of illicit medicines, which included "two cases of quinine," sold for over $5,000. When a group of concerned northern physicians attending the 1864 meeting of the American Medical Association tried to put a stop to such practices by introducing a motion to lift the embargo on medicines, their proposal was immediately quashed.[13]

As the blockade tightened and an increasing number of Confederate ports came under federal control, smugglers found it impossible to keep up with the South's high demand for medicines. Southerners were thus forced to employ a variety of largely ineffective substitutes to treat their infirmities. The Confederacy's surgeon general, Samuel P. Moore, had recognized early in the war that the new nation would need to rely on its own natural resources for pharmaceuticals. To this end he established a string of drug laboratories throughout the South and in the spring of 1862 circulated a pamphlet to his surgeons containing a list of indigenous plants that could be used in place of specific medicines. Months earlier Moore had commissioned a South Carolina physician named Francis Porcher to compile a more thorough work containing similar information. The result of Porcher's research was a dense guidebook entitled *Resources of the Southern Fields and Forests,* which was published in 1863 and listed a number of quinine substitutes among other natural medicines. The bark of yellow poplar, chestnut, hazel alder, black alder, holly, Georgia bark, and dogwood trees were just a few of the malaria treatments the book recommended. The dogwood bark remedy proved especially popular, and Porcher reprinted a letter written by one of his Charleston colleagues claiming that Huguenot settlers in South Carolina were the first to use the tree to treat malarial fevers. Cottonseed tea, knotgrass, and thoroughwort (boneset) were also endorsed as ague elixirs. As shortages became acute at the end of 1863, Moore sent out his own recommendation for a quinine substitute consisting of 30 percent dogwood bark, an equal portion of poplar bark, and 40 percent willow bark mixed with whiskey. Several of his subordinates at the Pettigrew Hospital in Raleigh thought that rubbing turpentine on malaria patients' abdomens prevented paroxysms. The southern public had its own ideas. Newspaper editors received letters from readers who claimed knowledge of formulas that worked as well as quinine. Among the harshest of them was a tea that contained red pepper.[14]

Any relief that southerners experienced after imbibing these concoctions was purely psychosomatic. In reality only quinine could effectively control the symptoms of malaria, and the South's supply dwindled each year of the war. Even captured Union supplies—an important source of the medicine—became unreliable as the Confederacy's battlefield losses mounted. Southerners quickly realized that the botanical cures recommended by Richmond were useless. One Florida physician complained that the native remedies he tried did nothing for his fever patients. Another surgeon, W. T. Grant, admitted in the summer of 1864 that the South's search for a quinine substitute had failed. He held out hope, however, that either sodium chloride or turpentine might provide a cure.

The paucity of quinine and the widespread use of ineffective substitutes in the South allowed the region's malarial fevers to flourish. As a consequence, Confederate soldiers and civilians suffered immensely, especially during the sickly season, when mosquito activity was at its peak. Wartime travel restrictions meant that white southerners accustomed to fleeing their plantations for healthier climes during the summer and fall months were forced to stay home. The limited amount of quinine which made it through the blockade was quickly snatched up by the Confederate government, leaving civilians to fend for themselves. Those who could afford it paid exorbitant prices on the black market for the drug, while others wrote to relatives in the North requesting that it be sent by mail. "My precious mother," began Anna DeWolf Middleton from Summerville, South Carolina, in September 1864. "I write now to beg you to send in your next letter a quarter of an ounce of quinine. You know, in this climate, life depends upon quinine—and though large quantities come in every ship, it is taken up so immediately for the army that it is exceedingly difficult for private individuals to procure it even at a very high price." Middleton went on to express her concerns about what the quinine shortage would mean for her children. They would certainly suffer, she noted, if unlucky enough to contract "the fever." Her plight was hardly unique. The lack of quinine and resulting spike in malarial fevers created additional hardships for southerners, who were already grappling with other painful changes wrought by the war, such as food shortages and a breakdown of plantation discipline. The Union blockade not only prevented medicines from reaching the Confederacy but also kept southerners from escaping during the mosquito season, which aided the spread of plasmodium parasites.[15]

Soldiers in the Army of Northern Virginia noticed the change. They were in desperate need of quinine at a time when Surgeon General Moore could do little besides recommend substitutes for the drug. During the siege of Petersburg malaria sickened hundreds of Confederates tasked with disrupting Benjamin Butler's attempt to build a canal across Dutch Gap on the James River. Frustrated with enemy batteries at Trent Reach which were preventing Union ships from moving on Richmond, Butler hoped he could bypass the defenses altogether by digging a 174-yard-long ditch through a spit of land created by a bend in the river. The Confederates put together a joint army-navy operation to harass Butler, but malaria hindered its effectiveness. Rebels in artillery battalions positioned near the river were attacked by "malarial fevers of every type" but because of widespread shortages were not administered quinine. At the same time, the Confederacy's James River Squadron was "very much impaired" by a disease it had picked up while anchored near a marsh along the river. Three ironclads and three gunboats were eventually sent into battle, but the men on board these vessels also fell ill. A report issued by the fleet's commander on August 22, nine days after the skirmish, shows that five of his ships had more crewman in the naval hospital than were fit for duty. Malaria continued to plague the James River Squadron throughout the fall of 1864 and helped limit its effectiveness at a time when Lee desperately needed the Confederate navy to relieve pressure on his bleeding army, which was itself suffering from the disease.[16]

Grant's forces were also afflicted with malaria but were better able to deal with the problem. In July 1864, 5 percent of his army was sick with the disease, and the figure rose to 6 percent in August. The number of "quotidian intermittent fever" cases spiked in October and November. Malaria occurred at roughly the same rate in the Army of the Potomac in 1862 and 1864, but Grant's medical staff did a better job of controlling the disease than McClellan's had two years earlier. Quinine-laced whiskey was administered to troops on a daily basis, and mosquito nets were hung over hospital beds (a decision intended to keep pests away but one that unwittingly prevented the transmission of plasmodium parasites). Union troops in 1864 were also better fed and clothed than southerners, which made them less susceptible to a variety of diseases, including malaria. And unlike his predecessors, Grant did not let concerns about southern pestilence keep him from relentlessly pursuing the Confederate enemy. Malaria or no malaria, his forces achieved victory by following Lee's army wherever it went. Grant's tenacity during the late summer and fall of 1864 is especially impressive when one considers the timidity of his predecessors, who severely limited or postponed Union military operations along the Eastern Seaboard each year when the South's climate became "a better defense than casemated forts."[17]

Ironically, the naval blockade that worsened the South's malaria problem simultaneously reduced the threat of yellow fever. Outbreaks during the prewar period were generally caused by an infected mosquito or passenger traveling from an area where the disease existed year-round, such as the West Indies. Once inside the United States, the virus was quickly spread by the local *Aedes aegypti* population. By restricting the flow of commercial traffic to the Confederacy in order to starve it into submission, the Union navy inadvertently reduced the possibility of an epidemic. Quarantine and sanitation policies enacted by Union commanders such as Benjamin Butler limited the possibility even further. But despite these measures, a number of yellow fever outbreaks struck the South, causing tremendous suffering for the civilians who endured them and adding to the region's reputation for unhealthiness. Fathers and mothers already concerned about sons in combat were forced to watch their neighbors perish in a manner that even nineteenth-century Americans, accustomed as they were to death on a daily basis, considered gruesome.

Survivors tried to regain a sense of normalcy as quickly as possible. Union supply ships soon reappeared in places such as Key West, Florida, and blockade-runners continued to steam into the few ports that remained in Confederate hands late in the war, such as Charleston, South Carolina, and Galveston, Texas. Life and the great struggle between North and South continued until the Confederacy finally collapsed in the spring of 1865. But during each of the previous four autumns, the threat of yellow fever had hung over the various southern port towns that ringed the Atlantic coast and Gulf of Mexico. While the bloodshed lasted, outbreaks surfaced in or near New Bern, Wilmington, Hilton Head, Charleston, Key West, Pensacola, New Orleans, Sabine, Galveston, Houston, Matagorda, Lavaca, and Brownsville. Residents who did not flee their homes prayed that the pestilence would pass them by or put their faith in specious reports issued by city officials claiming that the disease was fast disappearing. Accustomed to living in denial about the problems peculiar to their region, southerners were ready to believe the best about the health of their communities up until the last moment. Others, like the residents of New Orleans welcomed the arrival of the disease, hoping that it would strike down their enemies and help the Confederacy achieve victory. One man, with the tacit approval of the Richmond government, even tried to use fomite-filled trunks as bioweapons. His plan was scientifically flawed, but it reveals the lengths some southerners were willing to go to in order to win the war. It also shows the awful respect they had for a terrifying disease that had plagued their home communities for generations.

Southerners also respected the malarial fevers that became unmanageable once the Confederacy's supplies of medicine dried up. They paid exorbitant prices on the black market for quinine or sought relief from homegrown substitutes recommended by Richmond. Their enemies, on the other hand, were awash in the antimalarial drug by the end of the war. Federal troops swallowed more than a million ounces of quinine sulfate and other cinchona derivatives while operating in mosquito-infested regions such as the Mississippi Delta and Tidewater Virginia. Two large pharmaceutical companies headquartered in Philadelphia, Rosengarten and Sons and Powers and Weightman, were able to corner the market in quinine by keeping a steady stream of it flowing into Washington. Unable to build comparable factories, Confederates were forced to rely on a quinine smuggling network which employed both men and women as traffickers. Blockade-runners eased the shortage somewhat by auctioning off crates of the medicine smuggled in from Europe via the West Indies. But in the end these methods could not keep the South supplied with adequate amounts of quinine. The resulting increase in the number of malarial cases, coupled with terrifying yellow fever outbreaks, helped make life inside the short-lived Confederacy a miserable ordeal for most southerners.[18]

Epilogue

OR THE MAJORITY of those who lived through the Civil War, recollections of great battles such as Antietam, Gettysburg, and Shiloh overshadowed less pleasant memories of jaundiced and feverish young soldiers writhing in agony upon their hospital beds. Disease, like slavery, seemed oddly out of place in a narrative that was supposed to heal the nation's wounds by highlighting the brave deeds of its northern and southern heroes. In reality troops on both sides endured febrile sweats, shakes, and searing abdominal pain at least as often as they saw combat. The American Civil War, like every war that preceded it and others that would follow, was a pestilential nightmare.

Mosquitoes were a major reason why. The insects compromised soldiers' immune systems and did more to weaken northern forces sent into marginal theaters such as North Carolina and Arkansas than Rebel armies. At the same time, they established yellow fever zones along the South's shorelines which cowed commanders on both sides. Fortunately for the Union, northern leaders such as Ulysses S. Grant and Benjamin Butler refused to be intimidated by the South's miasmatic environment (or Rebel armies, for that matter) and embraced the best science of the day in order to minimize disease-related casualties. Their success must have surprised Confederate military planners such as Robert E. Lee and Jefferson Davis, who thought Dixie's unhealthiest areas were safe from attack during the sickly season.[1]

Mosquitoes also tormented a southern public that had never prepared itself for a long war of attrition. For most Confederate civilians life's essentials—food, clothing, and medicine—took precedence over lofty political principles and "stay-the-course" speeches delivered late in the war by unrepentant fire-eaters. Richmond's inability to provide quinine, the miracle drug of the mid-nineteenth century, gave southerners one more reason to doubt their leaders and the wisdom of secession.

Regrettably, the defeat of the Confederacy and the disbanding of the armies that had fought so savagely on battlefields across the South did not put an end to the suffering caused by malaria and yellow fever. In fact, for many Americans these diseases became bigger problems during Reconstruction than they had been over the course of the war. The removal of the naval blockade and the resumption of regular maritime trading through southern port cities produced a spate of yellow fever epidemics that claimed the lives of thousands of people. New Orleans, for example, which had experienced only a handful of "yellow jack" fatalities during the Union occupation, once again became an epicenter of disease and death. With ships arriving daily from the West Indies and a return to less stringent sanitation and quarantine policies, conditions were ripe for yellow fever, which in 1867 alone killed more than 3,000 of the city's residents.

Other southern towns faced similar horrors after the war. Galveston, Texas, lost 1,150 people to yellow fever the same year the disease struck New Orleans; an epidemic in Memphis, Tennessee, in 1873 sent close to 2,000 people to their graves; another thousand perished three years later in Savannah, Georgia. But the biggest epidemiological catastrophe of the period occurred in 1878 when yellow fever swept through the Mississippi River Valley like a scythe, harvesting between 16,000 and 20,000 souls. Congress responded by creating a National Board of Health the following year, but it could do little to control a disease that baffled even the best medical minds of the era. Consequently, southerners faced the grim prospect of additional outbreaks for another two decades, until army surgeon Walter Reed's experiments in Cuba confirmed that *Aedes aegypti* mosquitoes transmitted the yellow fever virus. In what was arguably a fitting denouement the South's last epidemic occurred in 1905 in New Orleans, which had witnessed more yellow fever deaths than any other city in the country.[2]

Malaria was also a major health issue after the war. Infected Union soldiers returning home from the battlefields of the South reintroduced the disease into northern communities, such as the Connecticut River Valley, which had not dealt with it in decades. Between 1870 and 1890 the number of new infections in New England exploded. Immigrants contributed to this epidemic by bringing new strains of malaria to the Northeast from southern and eastern Europe.[3]

Across the South poverty, ignorance, and the environmental and infrastructural damage wrought by the war allowed the region's plasmodium parasites to flourish well into the twentieth century. Authorities in every southern state reported dramatic increases in the frequency and intensity of outbreaks in the years immediately following Appomattox. The disease continued to plague the region until the federal government implemented a massive mosquito control program during World War II, which included the widespread use of the insecticide DDT.[4]

This program's success in controlling the mosquito population made it easy for subsequent generations of Americans to forget about the threat these insects had once posed to the health of their communities. Yellow fever and malaria became ailments associated with "developing countries" (a label that describes the antebellum South as accurately as it does any burgeoning democracy in modern-day Asia, Africa, or South America). Popular histories of the war perpetuated the public's amnesia by failing to include detailed discussions of soldiers' diseases. A handful of scholars and curious physicians tried to pick up the slack, but their publications gained little traction outside of classrooms and academic conferences.

And yet the diseases that were still medical mysteries in the nineteenth century were inextricably linked to the successes and failures of the armies that fought the war. Epidemiological history may never receive the same popular acclaim given to biographies and crisp military narratives, but it demands the attention of those who are dedicated to understanding the messy realities of life in previous centuries. L. P. Hartley's keen observation that "the past is a foreign country" is true from a microbiological standpoint as well as a cultural or political one.

Disease studies such as this one help us separate historical fact from the myths created by poets and filmmakers, whose success often hinges on their ability to make us nostalgic for simpler times dominated by larger-than-life heroes. Gruesome deaths caused by nasty viruses or bacterial infections are almost always missing from their Iliadic accounts of the war. Director John Huston's big-screen adaptation of the classic novel *The Red Badge of Courage*, for example, shows actor Audie Murphy (Henry Fleming) enduring all of the horrors of war except for disease. Ironically, Murphy had contracted malaria during World War II when he was a real-life American soldier fighting in Italy. Another, more recent Civil War movie that has enjoyed greater commercial success, *Glory* portrays black members of the Fifty-fourth Massachusetts Infantry battling racism and Rebels but not rheumatism; in reality northern surgeons diagnosed twenty-five thousand cases of that ailment among actual Afro-Yankee regiments. These are just two examples of how art can sometimes distort our view of the past even as it inspires and entertains us. From Hollywood's perspective, of course, epidemic disease is one of those inconvenient (and therefore expendable) facts of history which has the potential to spoil a good story.[5]

Military historians, however, ignore disease at their peril. For in truth a sick soldier is just as useless to his commander as one who has been wounded or killed by enemy fire. By tearing down the barriers that separate medical science from military studies, scholars strengthen their chances of finding new ways of explaining why certain battles and campaigns turned out the way they did. This is especially important for Civil War historians, who toil in a field of inquiry that has been plowed over many times by both professional antiquarians and members of the general public. Epidemiological history offers a seemingly limitless supply of new research possibilities. This work has focused on how mosquito-borne illness made a difference in some of the war's biggest campaigns, but the full extent to which other maladies affected these events remains a mystery. Diarrhea and dysentery, for example, the two most frequently diagnosed "diseases" of the war (symptoms really), sickened hundreds of thousands of soldiers on both sides, leaving historians to ponder what might have happened if even half of these men had stayed healthy. The connection between epidemiology and the military history of the conflict then deserves the same level of serious academic inquiry normally reserved for more conventional topics, such as the tactical decisions made by commanders.

By unwittingly serving as soldiers, mosquitoes have done more to shape our history than most people realize. When Napoléon Bonaparte attempted to quell a slave revolt on St. Domingue and reestablish a French colonial presence in North America, *Aedes aegypti* mosquitoes sickened thousands of his unacclimated European soldiers, ensuring the survival of the first black republic in the Western Hemisphere and prompting the Corsican emperor to sell Louisiana to the United States. Later in the same century anopheles mosquitoes helped Seminole Indians resist the federal government long after other tribes in the East had accepted Washington's authority. Troops sent into Florida's swamps to subdue the recalcitrant Native American nation were decimated by malarial fevers. During the Spanish-American War far more American soldiers were prostrated by Cuba's mosquitoes than were wounded in combat. The scientific discoveries that emerged from this conflict as a result of the army's yellow fever experiments were applied during the building of the Panama Canal, allowing American construction teams to avoid the mosquito-borne illnesses that had caused major setbacks for their French predecessors. During World War II malaria-carrying mosquitoes sickened U.S. soldiers serving in the Philippines and other regions. In the jungles of Vietnam American GIs encountered drug-resistant strains of the disease, sparking the rapid development of new medicines at the Walter Reed Army Institute of Research in Washington, D.C. More recently, American troops serving with the International Security Assistance Force in Afghanistan have been battling plasmodial infections as well as Taliban guerrillas.

None of this would have surprised Aldo Leopold. In his landmark study *A Sand County Almanac* the twentieth-century environmentalist argued that history cannot be interpreted solely in terms of human agency because humanity is in a symbiotic relationship with the environment. The mosquito's role in the events of the Civil War proves that he was right.[6]

APPENDIX 1

INCIDENCE OF MOSQUITO-BORNE DISEASE, 1861–1865

Note: The maps of malarial incidence among Union troops are based on data from Joseph Janvier Woodward, *The Medical and Surgical History of the War of the Rebellion. Part II. Volume I. Medical History* (Washington, D.C.: Government Printing Office, 1870). The map of yellow fever outbreaks is based on data from various primary and secondary sources.

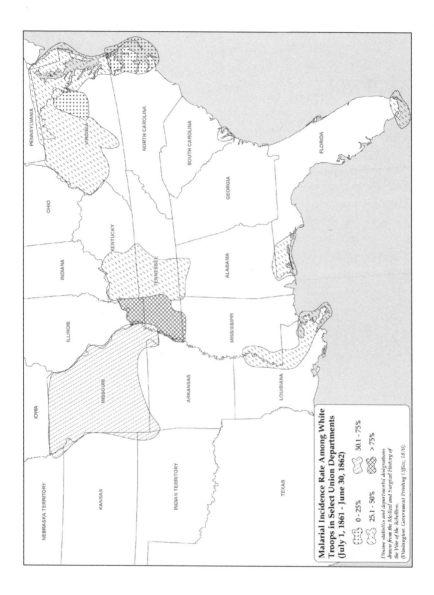

Malarial Incidence Rate Among White Troops in Select Union Departments (July 1, 1861 - June 30, 1862)

0 - 25%   50.1 - 75%
25.1 - 50%   > 75%

Disease statistics and departmental designations drawn from the Medical and Surgical History of the War of the Rebellion. (Washington: Government Printing Office, 1870).

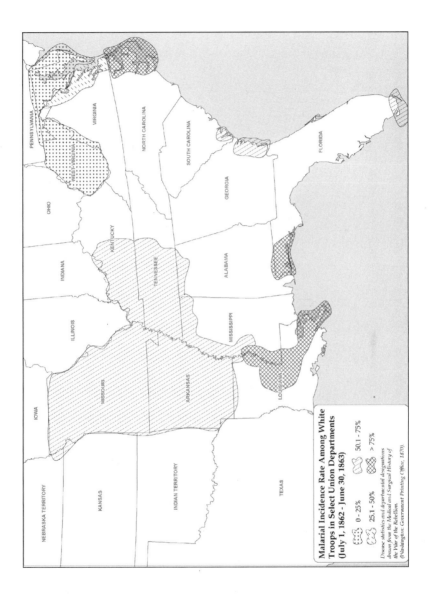

Malarial Incidence Rate Among White Troops in Select Union Departments (July 1, 1862 - June 30, 1863)

0 - 25%   50.1 - 75%

25.1 - 50%   >75%

Disease statistics and departmental designations drawn from the Medical and Surgical History of the War of the Rebellion. (Washington: Government Printing Office, 1870).

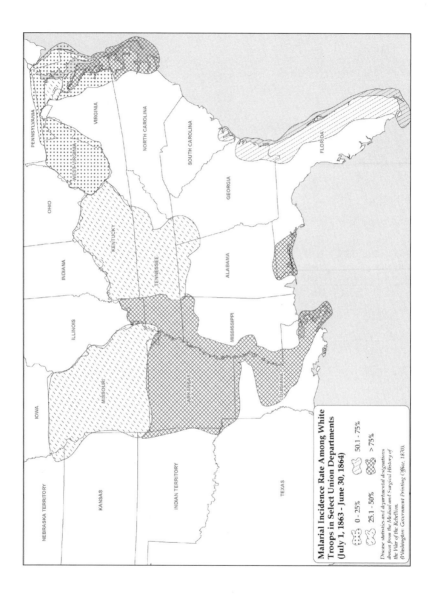

Malarial Incidence Rate Among White Troops in Select Union Departments (July 1, 1863 - June 30, 1864)

0 - 25%   50.1 - 75%

25.1 - 50%   > 75%

Disease statistics and departmental designations drawn from the Medical and Surgical History of the War of the Rebellion. (Washington: Government Printing Office, 1870).

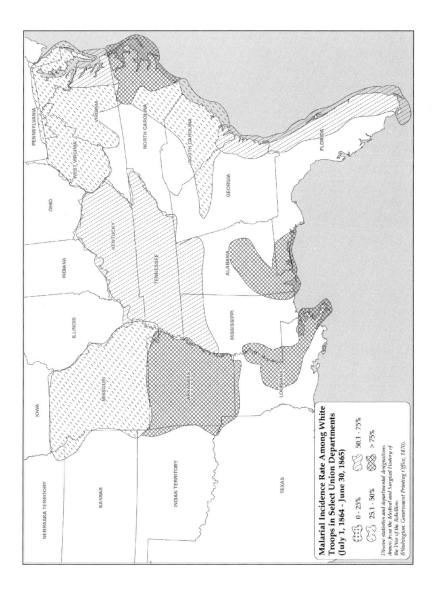

Malarial Incidence Rate Among White Troops in Select Union Departments (July 1, 1864 - June 30, 1865)

0 - 25%   50.1 - 75%

25.1 - 50%   >75%

Disease statistics and departmental designations drawn from the Medical and Surgical History of the War of the Rebellion. (Washington: Government Printing Office, 1870).

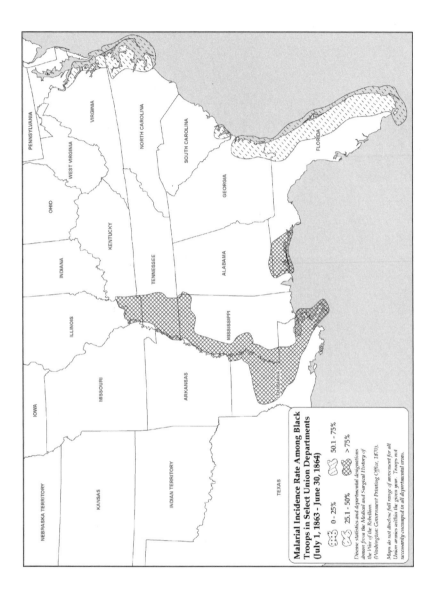

Malarial Incidence Rate Among Black Troops in Select Union Departments (July 1, 1863 – June 30, 1864)

Disease statistics and departmental designations drawn from the Medical and Surgical History of the War of the Rebellion. (Washington: Government Printing Office, 1870).

Maps do not disclose full range of movement for all Union armies within the given year. Troops not necessarily encamped in all departmental areas.

- 0 – 25%
- 25.1 – 50%
- 50.1 – 75%
- >75%

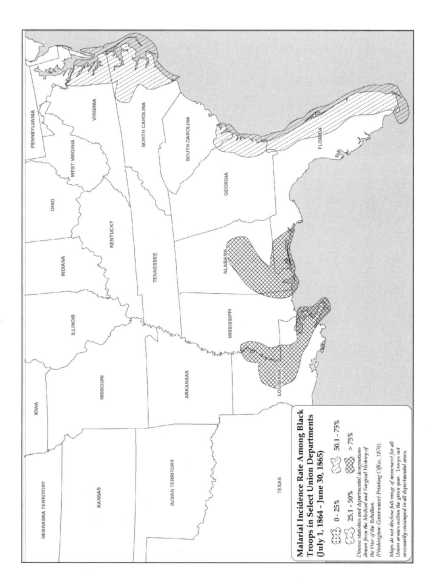

Malarial Incidence Rate Among Black Troops in Select Union Departments (July 1, 1864 - June 30, 1865)

0 - 25%   50.1 - 75%

25.1 - 50%   > 75%

Disease statistics and departmental designations drawn from the Medical and Surgical History of the War of the Rebellion. (Washington: Government Printing Office, 1870).

Maps do not disclose full range of movement for all Union armies within the given year. Troops not necessarily encamped in all departmental areas.

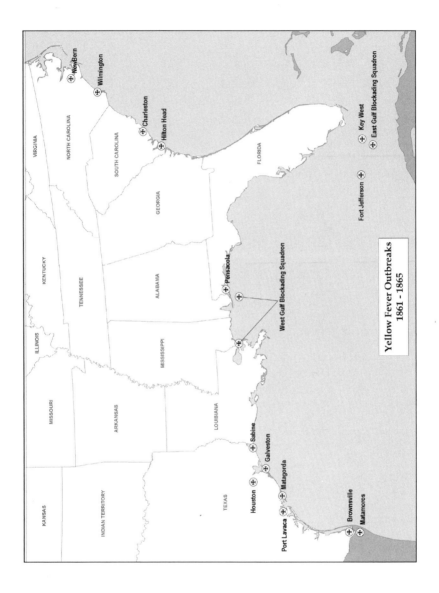

Yellow Fever Outbreaks
1861 - 1865

APPENDIX 2

COMMON DIAGNOSES AMONG UNION TROOPS, 1861–1866

DISEASE	NUMBER OF CASES
Diarrhea/dysentery	1,617,882[a]
Malaria	1,315,955[b]
Rheumatism (joint pain)	286,863
Catarrh	229,943
Bronchitis	221,008
Constipation	163,164
Gonorrhea	102,893
Colic (abdominal pain)	84,498
Boils (skin infection)	83,170
Headache	81,594
Syphilis	79,589
Typhoid fever	79,462
Inflammation of lungs	77,335
Jaundice (yellowing of skin, eyes)	77,236
Measles	76,318
Inflammation of conjunctiva (infection of eye membrane)	70,892
Inflammation of tonsils	66,665
Neuralgia (nerve pain)	64,792
Piles (hemorrhoids)	60,958
Mumps	60,314
Abscess (skin infection)	49,622
Scurvy	46,931

Dyspepsia (upset stomach)	40,121
Inflammation of pleura (infection of lung membrane)	39,027
Itch (various skin disorders)	35,236
Hernia	26,566
Erysipelas (skin infection)	24,812
Anemia	24,663
Consumption/scrofula (tuberculosis)	23,360
Inflammation of liver	1,329
Inflammation of larynx (throat infection)	20,056
Smallpox/varioloid	8,952

[a] There were 1,330,360 of diarrhea and 287,522 of dysentery.

[b] Excludes cases of "typho-malarial fever."

Source: The Medical and Surgical History of the War of the Rebellion (Washington, D.C.: Government Printing Office, 1870).

APPENDIX 3

Civil War Chronology

1861

April 12—Confederates shell Fort Sumter (S.C.)

April 19—Sixth Massachusetts Volunteer Infantry attacked by pro-secessionist mob in Baltimore, Md.

July 21—Battle of First Manassas (Va.)

August 6—Congress approves First Confiscation Act

August 10—Battle of Wilson's Creek (Mo.)

August 28—U.S. forces capture Fort Hatteras on the Outer Banks (N.C.)

September 6—U.S. forces capture Paducah (Ky.)

September 11–16—Cheat Mountain Campaign (present-day W.Va.)

September 12–20—Siege of Lexington (Mo.)

October 21—Battle of Ball's Bluff (Va.)

November 7—Battle of Port Royal Sound (S.C.) gives Federals control over Hilton Head and Beaufort; Battle of Belmont (Mo.)

1862

February 6—Grant captures Fort Henry (Tenn.)

February 8—Battle of Roanoke Island (N.C.)

February 16—Grant captures Fort Donelson (Tenn.)

February 25—U.S. forces occupy Nashville (Tenn.)

March 6–8—Battle of Pea Ridge (Ark.)

March 14—U.S. forces capture New Berne (N.C.) and New Madrid (Mo.)

April 5—Siege of Yorktown begins (Va.)

April 6–7—Battle of Shiloh (Tenn.); John Pope captures Island No. 10 (Mississippi River) on April 7

April 11—U.S. forces capture Fort Pulaski on the Savannah River (Ga.) and occupy Huntsville (Ala.)

April 25—U.S. forces capture New Orleans (La.) and Fort Macon (near Beaufort, N.C.)

May 5—Battle of Williamsburg (Va.)

May 8—Battle of McDowell (Va.)

May 25—Battle of Winchester (Va.)

May 30—Confederates evacuate Corinth (Miss.)

May 31–June 1—Battle of Seven Pines (Va.)

June 6—Battle of Memphis (Tenn.)

June 8—Battle of Cross Keys (Va.)

June 9—Battle of Port Republic (Va.)

June 16—Battle of Secessionville (S.C.)

June 25—Seven Days Campaign begins (part of McClellan's Peninsular Campaign) (Va.)

June 26—Battle of Mechanicsville (Va.) (Seven Days battles / Peninsular Campaign)

June 27—Battle of Gaines's Mill (Seven Days battles / Peninsular Campaign)

June 29—Battle of Savage Station (Va.) (Seven Days battles / Peninsular Campaign)

June 30—Battle of White Oak Swamp (Seven Days battles / Peninsular Campaign)

July 1—Battle of Malvern Hill (Seven Days battles / Peninsular Campaign)

July 17—Lincoln approves Second Confiscation Act

July 22—Lincoln announces his plan for emancipation to his cabinet

August 3—Halleck orders McClellan to withdraw from the Peninsula (Va.)

August 5—Battle of Baton Rouge (La.)

August 9—Battle of Cedar Mountain (Va.)

August 29–30—Battle of Second Manassas (Va.)

September 14—Battle of South Mountain (Md.)

September 15—Jackson captures Harper's Ferry (present-day W.Va.)

September 17—Battle of Antietam (Md.)

September 19—Battle of Iuka (Miss.)

September 22—Lincoln announces his plan for emancipation to the public

October 3–4—Battle of Corinth (Miss.)

October 8—Battle of Perryville (Ky.)

December 7—Battle of Prairie Grove (Ark.)

December 13—Battle of Fredericksburg (Va.)

December 20—Van Dorn captures Grant's supplies at Holly Springs (Miss.)

December 29—Battle of Chickasaw Bayou (Miss.)

December 31–January 2, 1863—Battle of Stone's River (Tenn.)

1863

January 1—Emancipation Proclamation goes into effect

January 11—U.S. forces capture Arkansas Post

January 20–22—Burnside's "Mud March" (Va.)
March 3—Lincoln signs Draft Bill
March 17—Battle of Kelly's Ford (Va.)
April 2—Bread Riot in Richmond (Va.)
April 7—Flag Officer Samuel Du Pont attacks Charleston (S.C.)
April 16—Federal fleet floats past Vicksburg (Miss.)
April 17—Grierson's Raid begins (Miss.)
May 1–4—Battle of Chancellorsville (Va.); Battle of Port Gibson (Miss.)
on May 1 (part of Grant's Vicksburg Campaign)
May 10—Stonewall Jackson dies
May 12—Battle of Raymond (Miss.) (Vicksburg Campaign)
May 14—Battle of Jackson (Miss.) (Vicksburg Campaign)
May 16—Battle of Champion's Hill (Miss.) (Vicksburg Campaign)
May 17—Battle of Big Black River Bridge (Miss.) (Vicksburg Campaign)
May 18—Vicksburg siege begins (Miss.)
May 21—Port Hudson siege begins (La.)
June 9—Battle of Brandy Station (Va.)
June 14–15—Battle of Second Winchester (Va.)
July 1–3—Battle of Gettysburg (Pa.)
July 4—Confederates surrender Vicksburg (Miss.); Lee withdraws
from Gettysburg (Pa.)
July 8—Confederates surrender Port Hudson (La.)
July 11–13—New York City draft riots
September 10—U.S. forces capture Little Rock (Ark.)
September 19–20—Battle of Chickamauga (Ga.)
November 19—Lincoln delivers his Gettysburg Address (Pa.)
November 24—Battle of Lookout Mountain (Tenn.)
November 25—Battle of Missionary Ridge (Tenn.)
November 26—Meade's Mine Run Campaign begins (Va.)

1864

February 14—U.S. forces capture Meridian (Miss.)
February 20—Battle of Olustee (Fla.)
March 1—Federal cavalry raid on Richmond fails (Va.)
March 9—Grant is promoted to lieutenant general
March 11—Red River Campaign begins (La.)
April 8—Battle of Sabine Crossroads (La.)
April 9—Battle of Pleasant Hill (La.)
April 12—Fort Pillow captured by Confederates (Tenn.)
May 5–6—Battle of the Wilderness (Va.)
May 7—Sherman's Atlanta Campaign begins (Ga.)

May 8–19—Battle of Spotsylvania Court House (Va.)

May 11—Battle of Yellow Tavern (Va.)

May 14–15—Battle of Resaca (Ga.) (Atlanta Campaign); Battle of New
Market (Va.) on May 15

May 16—Battle of Drewry's Bluff (Va.)

May 23–26—Battle of North Anna (Va.)

May 31–June 3—Battle of Cold Harbor (Va.)

June 14—Battle of Pine Mountain (Atlanta Campaign) (Ga.)

June 18—Petersburg siege begins (Va.)

June 27—Battle of Kennesaw Mountain (Atlanta Campaign) (Ga.)

July 9—Battle of Monocacy (Md.)

July 14—Battle of Tupelo (Miss.)

July 20—Battle of Peachtree Creek (Atlanta Campaign) (Ga.)

July 22—Battle of Atlanta (Atlanta Campaign) (Ga.)

July 28—Battle of Ezra Church (Atlanta Campaign) (Ga.)

July 30—Battle of the Crater (Va.); Confederates burn Chambersburg (Pa.)

August 5–8—Battle of Mobile Bay (Ala.)

August 18–19—Battle of Weldon Railroad (Va.)

August 25—Battle of Reams's Station (Va.)

August 31–September 1—Battle of Jonesboro (Atlanta Campaign) (Ga.)

September 2—U.S. forces occupy Atlanta

September 19—Third Battle of Winchester (Va.)

September 22—Battle of Fisher's Hill (Va.)

September 29–October 2—Battle of Peebles's Farm; Battle of Fort Harrison
on September 29–30 (Va.)

October 19—Battle of Cedar Creek (Va.)

October 23—Battle of Westport (Mo.)

November 8—Lincoln wins second term

November 16—Sherman's "March to the Sea" begins

November 21—Hood's March to Tennessee begins

November 30—Battle of Franklin (Tenn.)

December 15–16—Battle of Nashville (Tenn.)

December 20—Sherman captures Savannah (Ga.)

December 24—U.S. forces bombard Fort Fisher near Wilmington (N.C.)

1865

January 15—U.S. forces capture Fort Fisher (N.C.)

January 19—Sherman's March through the Carolinas begins

February 5–7—Battle of Hatcher's Run (Petersburg siege) (Va.)

February 17—U.S. forces occupy Columbia (S.C.); Confederates evacuate
Charleston (S.C.)

February 22—U.S. forces occupy Wilmington (N.C.)

March 8–10—Battle of Kinston (N.C.)

March 16—Battle of Averasboro (N.C.)

March 19–21—Battle of Bentonville (N.C.)

April 1—Battle of Five Forks (Va.)

April 3—U.S. forces occupy Richmond (Va.)

April 6—Battle of Sayler's Creek (Va.)

April 9—Lee surrenders at Appomattox Court House (Va.)

April 12—Confederates surrender Mobile (Ala.)

April 14—Lincoln shot at Ford's Theatre (Washington, D.C.)

April 26—Johnston surrenders at Durham Station (N.C.)

May 4—Taylor Surrenders at Citronelle (Ala.)

May 26—Confederate Army of the Trans-Mississippi surrenders at New Orleans (La.)

NOTES

INTRODUCTION

1. Mark F. Boyd, "An Historical Sketch of the Prevalence of Malaria in North America," *American Journal of Tropical Medicine* 21 (March 1941): 234; H. H. Cunningham, *Doctors in Gray: The Confederate Medical Service* (Baton Rouge: Louisiana State University Press, 1958), 191; Joseph Janvier Woodward, *The Medical and Surgical History of the War of the Rebellion*, pt. 2, vol. 1: *Medical History* (Washington, D.C.: Government Printing Office, 1879), 1–2, 401–2, 799–800; George Worthington Adams, *Doctors in Blue: The Medical History of the Union Army in the Civil War* (Baton Rouge: Louisiana State University Press, 1952), 202.

2. Charles E. Rosenberg, *Explaining Epidemics and Other Studies in the History of Medicine* (Cambridge: Cambridge University Press, 1992), 12–13; John Harley Warner, *The Therapeutic Perspective: Medical Practice, Knowledge, and Identity in America, 1820–1885* (Princeton: Princeton University Press, 1997), viii–x, 91.

3. Alfred Jay Bollet, *Civil War Medicine: Challenges and Triumphs* (Tucson: Galen Press, 2002), 58–59.

4. Warner, *Therapeutic Perspective*, 38.

5. Warner, *Therapeutic Perspective*, 87 (quotation).

6. Dale C. Smith, "The Rise and Fall of Typhomalarial Fever: 1. Origins," *Journal of the History of Medicine and Allied Sciences* 37 (1982): 182–220.

7. William H. McNeill, *Plagues and Peoples* (Garden City, N.Y.: Anchor Books, 1976); Elizabeth A. Fenn, *Pox Americana: The Great Smallpox Epidemic of 1775–82* (New York: Hill & Wang, 2001).

8. James I. Robertson Jr., *Soldiers Blue and Gray* (Columbia: University of South Carolina Press, 1988).

CHAPTER 1: AEDES, ANOPHELES, AND THE SCOURGES OF THE SOUTH

1. Hudson Strode, *Jefferson Davis: American Patriot, 1808–1861* (New York: Harcourt, Brace, 1955), 79–104; Haskell M. Monroe Jr. and James T. McIntosh, *The Papers of Jefferson Davis*, vol. 1: *1808–1840* (Baton Rouge: Louisiana State University Press, 1971), 407–32 (quotation); Zachary Taylor to General Jesup Thomas Sidney, December 15, 1820, Zachary Taylor Papers, 1820–1850, Manuscripts Division, Library of Congress; Trist Wood, "Jefferson Davis' First Marriage," *New Orleans Daily Picayune*, August 28, 1910; Varina Davis, *Jefferson Davis, Ex-President of the Confederate States of America: A Memoir by His Wife* (New York: Belford Co., 1890), 165–71.

2. Karen Ordahl Kupperman, "Fear of Hot Climates in the Anglo-American Colonial Experience," *William and Mary Quarterly* 41 (April 1984): 213–40; Daniel Drake, *A Systematic Treatise, Historical, Etiological, and Practical, on the Principal Diseases of the Interior Valley of North America, as They Appear in the Caucasian, African, Indian, and Esquimaux Varieties of Its Population*, ed. Norman D. Levine (Urbana: University of Illinois Press, 1964), 703–11; J. D. Rumph, "Thoughts on Malaria, and the Causes Generally of Fever," *Charleston Medical Journal and Review* 9 (July 1854): 439–46; William Webb, "On the So-Called Malarious Diseases. A Paper Read before the St. Louis Medical Society," *St. Louis Medical and Surgical Journal* 12 (December 1854): 481–85.

3. Perfect diagnoses in the absence of laboratory tests are of course impossible, but the location, time of year, and symptoms make it highly probable that the Davises were suffering from *Plasmodium falciparum*. Robert E. Sinden and Herbert M. Gilles, "The Malaria Parasites," in *Essential Malariology*, 4th ed., ed. David A. Warrell and Herbert M. Gilles (London: Arnold, 2002), 8–13; Hisashi Fujioka and Masamichi Aikawa, "The Malaria Parasite and Its Life-Cycle," in *Malaria: Molecular and Clinical Aspects*, ed. Mats Wahlgren and Peter Perlmann (Amsterdam: Overseas Publishers Association, 1999), 19–20; S. F. Kitchen, "The Infection in the Intermediate Host: Symptomatology, Falciparum Malaria," in *A Symposium on Human Malaria with Special Reference to North America and the Caribbean Region*, ed. Forest Ray Moulton (Washington, D.C.: American Association for the Advancement of Science, 1941), 196–207; Carol A. Butcher, "Malaria: A Parasitic Disease," *American Association of Occupational Health Nurses Journal* 52 (July 2004): 302–9; Sharon Parmet, "Malaria," *Journal of the American Medical Association* 291, June 2, 2004, 2664.

4. Donald J. Krogstad, "Plasmodium Species (Malaria)," in *Principles and Practice of Infectious Diseases*, 5th ed., ed. Gerald L. Mandell, John E. Bennett, and Raphael Dolin (Philadelphia: Churchill Livingstone, 2000), 2817–29; Margaret Humphreys, *Malaria: Poverty, Race, and Public Health in the United States* (Baltimore: Johns Hopkins University Press, 2001), 9; Darrett B. Rutman and Anita H. Rutman, "Of Agues and Fevers: Malaria in the Early Chesapeake," *William and Mary Quarterly* 33 (January 1976): 31–60; Michael Colbourne, *Malaria in Africa* (London: Oxford University Press, 1966), 13–16; Richard F. Darsie Jr. and Ronald A. Ward, *Identification and Geographical Distribution of the Mosquitoes of North America, North of Mexico* (Gainesville: University Press of Florida, 2005), 286–97; N.R.H. Burgess and G. O. Cowan, *Atlas of Medical Entomology* (London: Chapman & Hall Medical, 1993), 11–14.

5. Elisha Bartlett, *The History, Diagnosis, and Treatment of the Fevers of the United States* (Philadelphia: Blanchard & Lea, 1852), 347–99; Edwin Samuel Gaillard, *An Essay on Intermittent and Bilious Remittent Fevers: With Their Pathological Relation to Ozone* (Charleston: Walker & Evans, Stationers and Printers, 1856), 17–18; Fr. Xavier DeRolette, *Lecture on the Fever & Ague and Other Intermittent Fevers* (Pittsburgh: A. A. Anderson & Sons, 1865), 5; Drake, *Treatise*, 703; A. P. Merrill, *Lectures on Fever, Delivered in the Memphis Medical College, in 1853–6* (New York: Harper & Bros., 1865), 98; John Duffy, "The Impact of Malaria on the South," in *Disease and Distinctiveness in the American South*, ed. Todd L. Savitt and James Harvey Young (Knoxville: University of Tennessee Press, 1988), 29–54.

6. Dale C. Smith, "The Rise and Fall of Typhomalarial Fever: 1. Origins," *Journal of the History of Medicine and Allied Sciences* 37 (1982): 182–220.

rainforest destruction is causing an explosion in the number of malaria-carrying mosquitoes. See Anne Underwood, "Tracking Disease," *Newsweek* 146, November 14, 2005, 46–48; Erwin H. Ackerknecht, *Malaria in the Upper Mississippi Valley, 1760–1900* (Baltimore: Johns Hopkins Press, 1945); Milo Milton Quaife, ed., *Growing Up with Southern Illinois, 1820 to 1861, from the Memoirs of Daniel Harmon Brush* (Chicago: Lakeside Press, 1944), 15–24 (quotation).

8. Charles Dickens, *American Notes for General Circulation*, ed. Patricia Ingham (London: Penguin, 2000), 190–91 (quotation); Samuel Thompson, "Malaria and Its Relation to the Existence and Character of Disease," *North-Western Medical and Surgical Journal* 3 (May 1850): 19–43; C. Handfield Jones, "Considerations Respecting the Operation of Malaria on the Human Body," *Association Medical Journal* 187 (August 1856): 668–70; W. H. Martin, "Various Forms of Intermittent Diseases," *Illinois and Indiana Medical and Surgical Journal* 2 (1847–48): 407–9; F. P. Leavenworth, "Malaria," *St. Louis Medical & Surgical Journal* 12 (January 1854): 35–58; Hiram Nance, "Retrospect of Miasmatic Diseases for a Few Years Past," *North-western Medical and Surgical Journal* 3 (February 1854): 54–57; J. R. Black, "On the Ultimate Cause of Malarial Disease," *St. Louis Medical and Surgical Journal* 12 (July 1854): 353–63.

9. Other diseases such as typhoid fever contributed to the fever season in the South and were frequently misdiagnosed as malaria by physicians. See Todd L. Savitt and James Harvey Young, eds., *Disease and Distinctiveness in the American South* (Knoxville: University of Tennessee Press, 1988); A. Cash Koeniger, "Climate and Southern Distinctiveness," *Journal of Southern History* 54 (February 1988): 21–44; Humphreys, *Malaria*, 11–13; Frederick Law Olmstead, *The Cotton Kingdom*, ed. Arthur M. Schlesinger Sr. (New York: Random House, 1984), 182–83 (quotation); Peter H. Wood, *Black Majority: Negroes in Colonial South Carolina from 1670 through the Stono Rebellion* (New York: W. W. Norton, 1974), 84–91. For additional information on African genetic immunity to malaria, see Christopher Warren, "Northern Chills, Southern Fevers: Race-Specific Mortality in American Cities, 1730–1900," *Journal of Southern History* 63 (February 1997): 23–56; and Kenneth F. Kiple, "A Survey of Recent Literature on the Biological Past of the Black," *Social Science History* 10 (Winter 1986): 343–67.

10. Bartlett, *History*, 388–89; Drake, *Treatise*, 704–11; Black, "Ultimate," 358–59; John Greenleaf Whittier, "The Farewell," quoted in Frederick Douglass, *Narrative of the Life of Frederick Douglass, an American Slave, Written by Himself*, with an introduction by Peter J. Gomes (New York: New American Library, 1997), 60.

11. "Address of Dr. S. S. Satchwell, on Malaria" in *Transactions of the Medical Society of the State of North Carolina, at Its Third Annual Meeting Held in Wilmington, N.C., May, 1852* (Wilmington: T. Loring, 1852), 39–62; Ackerknecht, *Malaria*, 19; Bartlett, *History*, 388; Edward Warren, "Observations on Miasmatic Diseases," *American Medical Monthly* 2 (October 1854): 241–53; Drake, *Treatise*, 713; Thompson, "Malaria," 20–21; Duffy, "Malaria," 38.

12. William Pepper, *On the Use of Bebeerine and Cinchonia in the Treatment of Intermittent Fever* (Philadelphia: T. K. and P. G. Collins, 1853), 8–14; James O. Breeden, "Disease as a Factor in Southern Distinctiveness," in Savitt, *Disease and Distinctiveness*, 9–11; Rene LaRoche, *Pneumonia: Its Supposed Connection, Pathological and Etiological, with Autumnal Fevers; Including an Inquiry into the Existence and Morbid Agency of Malaria*

(Philadelphia: Blanchard & Lea, 1854), 57, 407–8; Koeniger, "Climate," 35; David R. Goldfield, "The Business of Health Planning: Disease Prevention in the Old South," *Journal of Southern History* 42 (November 1976): 557–70; Richard D. Arnold, *An Essay upon the Relation of Bilious and Yellow Fever, Prepared at the Request of, and Read before the Medical Society of the State of Georgia, at Its Session Held at Macon* (Augusta: J. Morris, 1856), 4–5; William R. Horsfall, *Medical Entomology: Arthropods and Human Disease* (New York: Ronald Press Co., 1962), 340.

13. The last yellow fever outbreak to occur north of Virginia happened in New York City in 1822. John Duffy, "Yellow Fever in the Continental United States during the Nineteenth Century," *Bulletin of the New York Academy of Medicine* 44 (June 1968): 687–701; K. David Patterson, "Yellow Fever Epidemics and Mortality in the United States, 1693–1905," *Social Science and Medicine* 34 (1992): 855–65; Jo Ann Carrigan, "Yellow Fever: Scourge of the South," in Savitt, *Disease and Distinctiveness*, 55–78.

14. Margaret Humphreys, *Yellow Fever and the South* (Baltimore: Johns Hopkins University Press, 1992), 4–5; S. Rickard Christophers, *Aedes Aegypti: The Yellow Fever Mosquito* (Cambridge: Cambridge University Press, 1960), 57; Thomas M. Rivers, *Viral and Rickettsial Infections of Man*, 2nd ed., (Philadelphia: J. B. Lippincott Co., 1952), 538–44; Andrew Spielman and Michael D'Antonio, *Mosquito: A Natural History of Our Most Persistent and Deadly Foe* (New York: Hyperion, 2001), 54–61.

15. John Duffy, "Medical Practices in the Ante Bellum South," *Journal of Southern History* 25 (February 1959): 53–72; Patterson, "Yellow Fever," 858; *New Orleans Daily Picayune*, September 2, 3, and 7, 1853; Sheldon Watts, *Epidemics and History: Disease, Power, and Imperialism* (New Haven: Yale University Press, 1997), 243–44; Spielman, *Mosquito*, 63; William L. Robinson, *The Diary of a Samaritan. By a Member of the Howard Association of New Orleans* (New York: Harper & Brothers, 1860), 121, 259.

16. *Savannah Daily Morning News*, September 12, 19, 20, 28, and 29, 1854; *Charleston Daily Courier*, September 16, 1854.

17. D. J. Cain, *History of the Epidemic of Yellow Fever in Charleston, S.C., in 1854* (Philadelphia: T. K. & P. G. Collins, 1856), 7–9; *Charleston Courier*, September 14 and 15, 1854; *Savannah Morning News*, September 19, 1854; W. L. Felder, "Observations on the Yellow Fever Epidemic of 1854, in Augusta, Ga.," *Southern Medical and Surgical Journal* 11 (October 1855): 598–608; Patterson, "Yellow Fever," 858.

18. William McCraven, "The Epidemic Yellow Fever of 1859, in Houston, Texas," *New Orleans Medical News and Hospital Gazette* 7 (April 1860): 105–10; Ashbel Smith, "On the Climate, Etc., of a Portion of Texas," *Southern Medical Reports* 2 (1850): 453–59; *Houston Telegraph*, October 13, 1858.

19. *Report on the Origin of the Yellow Fever in Norfolk during the Summer of 1855. Made to the City Councils by a Committee of Physicians* (Richmond, Va.: Ritchie & Dunnavant, 1857), 20–38; George D. Armstrong, *The Summer of Pestilence: A History of the Ravages of the Yellow Fever in Norfolk, Virginia, A.D. 1855* (Philadelphia: J. B. Lippincott & Co., 1856), 94–100; *Report of the Portsmouth Relief Association to the Contributors of the Fund for the Relief of Portsmouth, Virginia, during the Prevalence of the Yellow Fever in That Town in 1855* (Richmond, Va.: H. K. Ellyson's Steam Power Presses, 1856), 129–30; Patterson, "Yellow Fever," 858.

20. *Report of the Philadelphia Relief Committee Appointed to Collect Funds for the Sufferers by Yellow Fever, at Norfolk and Portsmouth, Va., 1855* (Philadelphia: Inquirer

Moncure D. Conway, *The True and the False in Prevalent Theories of Divine Dispensations* (Washington, D.C.: Taylor & Maury, 1855), 17 (quotation); Carrigan, "Yellow Fever," 60–61; Richard S. Storrs, *Terrors of the Pestilence: A Sermon, Preached in the Church of the Pilgrims, Brooklyn, N.Y., on Occasion of a Collection in Aid of the Sufferers at Norfolk, Va., September 30th, 1855* (New York: John A. Gray, 1855), 20 (quotation).

21. Alexander F. Vaché, *Letters on Yellow Fever, Cholera, and Quarantine; Addressed to the Legislature of the State of New York* (New York: McSpedon & Baker, 1852), 7–17; Edward Jenner Coxe, *Practical Remarks on Yellow Fever, Having Special Reference to the Treatment* (New Orleans: J. C. Morgan, 1859), 9–10; Thomas Anderson, *Handbook for Yellow Fever: Describing Its Pathology and Treatment* (London: Churchill & Sons, 1856), 10–14; Thomas Y. Simons, *An Essay on the Yellow Fever, as It Has Occurred in Charleston, Including Its Origin and Progress up to the Present Time* (Charleston, S.C.: Steam Power–Press of Walker and James, 1851), 10–23; *New Orleans Daily Picayune*, September 4, 1853; Humphreys, *Yellow Fever*, 8; "The Annual Report of the Board of Health, for the Year 1850, as Required by Law," *New Orleans Medical and Surgical Journal* 8 (March 1851): 590–91.

22. Although modern-day historians disagree over whether or not African immigrants possessed genetic immunity to yellow fever, nineteenth-century physicians were convinced that slaves were immune to the disease. See Margaret Humphreys, *Intensely Human: The Health of the Black Soldier in the American Civil War* (Baltimore: Johns Hopkins University Press, 2008), 48–49; Carrigan, "Yellow Fever," 59–63; Humphreys, *Yellow Fever*, 6–7; H. R. Carter, *Yellow Fever: Its Epidemiology, Prevention, and Control* (Washington, D.C.: Government Printing Office, 1914), 14; Watts, *Epidemics*, 217; "Yellow Fever Vaccine," *World Health Organization Weekly Epidemiological Record* 78 (2003): 349–59; Samuel A. Cartwright, "Report on the Diseases and Physical Peculiarities of the Negro Race" in Arthur L. Caplan, H. Tristram Engelhardt Jr., and James L. McCartney, eds., *Concepts of Health and Disease: Interdisciplinary Perspectives* (Reading, Mass.: Addison-Wesley, 1981), 314; Kenneth F. Kiple, "Black Yellow Fever Immunities, Innate and Acquired, as Revealed in the American South," *Social Science History* 1 (Summer 1977): 419–36; Wood, *Majority*, 80–82.

23. *The War of the Rebellion: A Compilation of the Official Records of the Union and Confederate Armies*, vol. 51, ser. 1, pt. 1, 369–70, hereafter referred to as *OR*; Alfred Jay Bollet, *Civil War Medicine: Challenges and Triumphs* (Tucson: Galen Press, 2002), 289–90.

24. Montgomery Meigs Diary (copy), March–September 1861, in John G. Nicolay Papers, box 13, Manuscript Division, Library of Congress, Washington, D.C.

25. "Urban" yellow fever requires a substantial and concentrated human population to survive and therefore most frequently appeared in cities. Sylvatic, or "jungle," yellow fever exists among tree-dwelling mammals in tropical ecosystems and is transmitted by *Haemagogus* mosquitoes. Thomas T. Smiley, "The Yellow Fever at Port Royal, S.C.," *Boston Medical and Surgical Journal* 67, January 8, 1863, 449–68 (quotation); John Ordronaux, *Hints on the Preservation of Health in Armies. For the Use of Volunteer Officers and Soldiers* (1861; rpt., San Francisco: Norman Publishing, 1990), 85–86 (quotation); "Yellow Fever on the Southern Coast, and How to Avoid It," *New York Times*, May 18, 1862; "The Fate of Virginia," *London Review*, July 6, 1861 reprinted in *Living Age* 70, August 10, 1861, 374–75 (quotation); *OR*, vol. 51, ser. 1, pt. 2, 550; H. H. Cunningham, *Doctors in*

Gray: The Confederate Medical Service (Baton Rouge: Louisiana State University Press, 1958), 191; Mark F. Boyd, "An Historical Sketch of the Prevalence of Malaria in North America," *American Journal of Tropical Medicine* 21 (March 1941): 223–44.

CHAPTER 2: GLORY OF GANGRENE AND "GALLINIPPERS"

1. H. H. Cunningham, *Doctors in Gray: The Confederate Medical Service* (Baton Rouge: Louisiana State University Press, 1958), 5; Alfred Jay Bollet, *Civil War Medicine: Challenges and Triumphs* (Tucson: Galen Press, 2002), 283–375; George Worthington Adams, *Doctors in Blue: The Medical History of the Union Army in the Civil War* (Baton Rouge: Louisiana State University Press, 1952), 3. "Malingering," or feigning illness, was also common in Civil War armies.

2. These numbers include "typho-malarial," "intermittent," "remittent," and "congestive" fever. *The Medical and Surgical History of the Civil War*, vol. 5 (Wilmington, N.C.: Broadfoot Publishing Co., 1991), 77–78, hereafter referred to as *MSH*; Bollet, *Medicine*, 289; Cunningham, *Gray*, 191; H. H. Cunningham, "The Medical Service and Hospitals of the Southern Confederacy" (Ph.D. diss., University of North Carolina, 1952), 159; Bell I. Wiley, *The Life of Johnny Reb* (New York: Bobbs-Merrill, 1943), 253.

3. At the bottom of the medical corps' chain of command were regimental surgeons who were supported by assistant surgeons and private contractors known as "acting assistant surgeons." Each brigade, division, army, and department also had its own medical officer, all of them under the supervision of the surgeon general. Michael A. Cooke, "The Health of the Union Military in the District of Columbia, 1861–1865," *Military Affairs* 48 (October 1984): 194–99; *United States Sanitary Commission Records*, ser. 7: Statistical Bureau Archives, Camp Inspection Returns, 1861–64, Library of Congress, Washington, D.C.; Geo. B. Willson, "Army Ambulances—Cases in the Hospital of Richardson's Brigade," *Boston Medical & Surgical Journal* 65 (January 1862): 542–44 (quotation); Charles S. Tripler to General George B. McClellan, February 7, 1863, *The War of the Rebellion: A Compilation of the Official Records of the Union and Confederate Armies*, vol. 5, ser. 1, 76–93 (quotation), hereafter referred to as *OR*.

4. John Duffy, "Medical Practice in the Ante Bellum South," *Journal of Southern History* 25 (February 1959): 53–72 (quotation); Joseph Janvier Woodward, M.D., *Outlines of the Chief Camp Diseases of the United States Armies as Observed during the Present War* (1863; rpt., San Francisco: Norman Publishing, 1992), 14–17.

5. Bollet, *Medicine*, 9–12, 227; "Rules for Preserving the Health of the Soldier. Report of the U.S. Sanitary Commission," *Harper's Weekly*, August 24, 1861, 542; United States Sanitary Commission Records, ser. 7: Statistical Bureau Archives, Camp Inspection Returns, 1861–64, Library of Congress, Washington, D.C.

6. John Brinton, *Personal Memoirs of John H. Brinton, Major and Surgeon U.S.V., 1861–1865* (New York: Neale Publishing Co., 1914), 60–61; *United States Sanitary Commission Records*, ser. 7; Henry H. Wright, *A History of the Sixth Iowa Infantry* (Iowa City: State Historical Society of Iowa, 1923), 30–31 (quotation).

7. Paul F. Eve, "Answers to Certain Questions Propounded by Prof. Charles A. Lee, M.D., Agent of the United States Sanitary Commission, Relative to the Health, &c., of the Late Southern Army," *Nashville Journal of Medicine and Surgery* 1 (July 1866): 12–32 (quotation); Bollet, *Medicine*, 270–71; Edward Porter Alexander, *Fighting for the*

Confederacy: The Personal Recollections of General Edward Porter Alexander (Chapel Hill: University of North Carolina Press, 1989), 35–36 (quotation); Margaret Humphreys, *Malaria: Poverty, Race, and Public Health in the United States* (Baltimore: Johns Hopkins University Press, 2001),14; John Duffy, "The Impact of Malaria on the South," in *Disease and Distinctiveness in the American South,* ed. Todd L. Savitt and James Harvey Young (Knoxville: University of Tennessee Press, 1988), 29–54.

8. William T. Sherman to G. Mason Graham, March 1, 1860, in Walter L. Fleming, ed., *General W. T. Sherman as College President* (Cleveland: Arthur M. Clark Co., 1912), 183–84 (quotation); Benjamin Butler, *Private and Official Correspondence of Gen. Benjamin F. Butler during the Period of the Civil War,* vol. 2 (Norwood, Mass.: Plimpton Press, 1917), 112–13, 272 (quotation); Alfred L. Castleman, *The Army of the Potomac. Behind the Scenes.* (Milwaukee: Strickland & Co., 1863), 110; William C. Holton, *Cruise of the U.S. Flag-ship Hartford, 1862–1863* (1863; rpt., Tarrytown, N.Y.: W. Abbatt, 1922), 39.

9. Richard Everett Wood, "Port Town at War: Wilmington, North Carolina 1860–1865" (Ph.D. diss., Florida State University, 1976), 176; *OR,* vol. 15, ser. 1, 144–45, 815–17; Margaret Humphreys, *Yellow Fever and the South* (Baltimore: Johns Hopkins University Press, 1992), 6.

10. Wiley, *Johnny Reb,* 249; J. H. Johnson to "Dearest Wife," April 16, 1863, Jonathan Huntington Johnson, *The Letters and Diary of Captain Jonathan Huntington Johnson* (n.p.: Alden Chase Brett, 1961), 112 (quotation); Junius N. Bragg to "My Dear Wife," July 19, 1863, J. N. Bragg, *Letters of a Confederate Surgeon, 1861–65* (Camden, Ark.: Hurley Co., 1960), 158–62 (quotation); Terrence J. Winschel, ed., *The Civil War Diary of a Common Soldier: William Wiley of the 77th Illinois Infantry* (Baton Rouge: Louisiana State University Press, 2001), 130 (quotation); Michael B. Ballard, *Vicksburg: The Campaign That Opened the Mississippi* (Chapel Hill: University of North Carolina Press, 2004), 375 (quotation); T. H. Barton, *Autobiography of Dr. Thomas H. Barton, the Self-Made Physician of Syracuse, Ohio, Including a History of the Fourth Regt. West Va. Vol. Infy.* (Charleston: West Virginia Printing Co., 1890), 157–58 (quotation); Reid Mitchell, *Civil War Soldiers* (New York: Penguin Books, 1988), 95–96.

11. Albert Theodore Goodloe, *Confederate Echoes: A Voice from the South in the Days of Secession and of the Southern Confederacy* (Nashville: Smith & Lamar, 1907), 248–49; Henry Warren Howe, *Passages from the Life of Henry Warren Howe, Consisting of Diary and Letters Written during the Civil War, 1861–1865* (Lowell, Mass.: Courier-Citizen Co., Printers, 1899), 127 (quotation); Isaiah Price, *History of the Ninety-seventh Regiment, Pennsylvania Volunteer Infantry, during the War of the Rebellion, 1861–65, with Biographical Sketches* (Philadelphia: By the Author for the Subscribers, 1875), 127 (quotation); "Medical History of the Seventeenth Regiment Mass. Volunteers," *Boston Medical and Surgical Journal* 68 (February 1863): 136–41.

12. *MSH,* 5:78; Bell I. Wiley, *The Life of Billy Yank* (New York: Bobbs-Merrill, 1951), 133 (quotation); Cunningham, *Gray,* 190; *MSH,* 3:142, 180, 186 (quotations); E. Andrews to "Messrs Editors," August 9, 1862, in *Chicago Medical Examiner* 3 (August 1862): 479–84 (quotation).

13. Some historians have argued that quinine rationing had a negligible effect on the health of troops, but the existing evidence suggests otherwise. Soldiers who received regular prophylactic doses of the drug were healthier than those who did not. Also, by the mid-nineteenth century most professionally trained physicians knew how much

quinine was needed to check intermittent and remittent fevers. See Dale C. Smith, "Quinine and Fever: The Development of the Effective Dosage," *Journal of the History of Medicine and Allied Sciences* 31 (July 1976): 343–67; and Samuel Logan, "Prophylactic Effects of Quinine," *Confederate States Medical & Surgical Journal* 1 (June 1864): 81–83. During the first year of the Civil War the United States Sanitary Commission (USSC) published reports endorsing the use of quinine as a prophylactic. Unfortunately, the USSC was only an advisory body and could not enforce its recommendations. See *Report of a Committee Appointed by Resolution for the Sanitary Commission, to Prepare a Paper on the Use of Quinine as a Prophylactic against Malarious Diseases* (New York: Wm. C. Bryant & Co., 1861); and "Rules for Preserving the Health of the Soldier," *Harper's Weekly*, August 24, 1861.

Michael A. Flannery, *Civil War Pharmacy: A History of the Drugs, Drug Supply and Provision, and Therapeutics for the Union and Confederacy* (New York: Pharmaceutical Products Press, 2004), 143–44; Bollet, *Medicine*, 236–38; Samuel Logan, "Prophylactic Effects of Quinine," *Confederate States Medical & Surgical Journal* 1 (June 1864): 81–83; Stephen Rogers, *Quinine as a Prophylactic or Protective from Miasmatic Poisoning, A Preventative of Paroxysms of Miasmatic Diseases* (Albany, N.Y.: Steam Press of C. Van Benthuysen, 1862), 4; "On the External Application of Oil of Turpentine as a Substitute for Quinine in Intermittent Fever, with Reports of Cases," *Confederate States Medical & Surgical Journal* 1 (January 1864): 7–8; "A Startling Fact!" *Harper's Weekly*, September 17, 1864 (quotation); Surgeon Geo. Hammond to the Surgeon General's Office, May 2, 1863, *Letters and Endorsements Sent to Medical Officers, Sept. 1862–Sept. 1872*, 4:35, Records of the Office of the Surgeon General (Army), Record Group 112, National Archives, Washington, D.C.

14. *Letters and Endorsements Sent to Medical Officers, Sept. 1862–Sept. 1872*, 4:175, (quotation); Charles McGregor, *History of the Fifteenth Regiment, New Hampshire Volunteers, 1862–1863* (Concord, N.H.: Published by Order of the Fifteenth Regiment Association, 1900), 261 (quotation); James W. Wheaton, comp., *Surgeon on Horseback: The Missouri and Arkansas Journal and Letters of Dr. Charles Brackett of Rochester, Indiana, 1861–1863* (Carmel: Guild Press of Indiana, 1998), 177 (quotation); Minetta Altgelt Goyne, *Lone Star and Double Eagle: Civil War Letters of a German-Texas Family* (Fort Worth: Texas Christian University Press, 1982), 140.

15. Hospital Tickets & Case Papers, Mississippi Squadron, 1862–64, Record Group 52, National Archives (quotation); Hospital Tickets & Case Papers, Pinkney Hospital, Memphis, 1863, Record Group 52, National Archives, Washington, D.C. (quotation); *The Official Records of the Union and Confederate Navies in the War of the Rebellion*, vol. 10, ser. 1, 735–36, hereafter referred to as ORN.

16. Frank R. Freemon, *Gangrene and Glory: Medical Care during the American Civil War* (Chicago: University of Illinois Press, 1998), 135–36 (quotation); Richard D. Arnold, *An Essay upon the Relation of Bilious and Yellow Fever* (Augusta, Ga.: J. Morris, 1856), 7–8; Medical Journal *U.S.S. Colorado*, June 4, 1861, to February 5, 1864, Record Group 52, National Archives, Washington, D.C. (quotation); *MSH*, 5:681 (quotation); G.R.B. Horner, M.D., "Notice of the Yellow Fever as It Occurred in Key West and in the U.S. East Gulf Blockading Squadron, in 1862," *American Journal of the Medical Sciences* 92 (October 1863): 391–98.

17. *MSH*, 5:678–82 (quotation); Bollet, *Civil War*, 235; Stanley B. Weld, "A Connecti-

cut Surgeon in the Civil War: The Reminiscences of Dr. Nathan Mayer," *Journal of the History of Medicine and Allied Sciences* 19 (July 1964): 272–86; Joseph Janvier Woodward, M.D., *Outlines of the Chief Camp Diseases of the United States Armies as Observed during the Present War* (1863; rpt., San Francisco: Norman Publishing, 1992), 160–61 (quotation); Flannery, *Pharmacy*, 131.

18. Alfred Jay Bollet, M.D., *Plagues & Poxes: The Impact of Human History on Epidemic Disease* (New York: Demos, 2004), 37.

CHAPTER 3: MOSQUITO COASTS

1. Shelby Foote, *The Civil War, a Narrative: Fort Sumter to Perryville* (New York: Vintage Books, 1986), 369–71; James McPherson, *Battle Cry of Freedom: The Civil War Era* (New York: Ballantine Books, 1988), 419–20; David J. Eicher, *The Longest Night: A Military History of the Civil War* (New York: Simon & Schuster, 2001), 239–42; Gerald M. Capers Jr., "Confederates and Yankees in Occupied New Orleans, 1862–1865," *Journal of Southern History* 30 (November 1964): 405–26.

2. K. David Patterson, "Yellow Fever Epidemics and Mortality in the United States, 1693–1905," *Social Science and Medicine* 34 (1992): 855–65; John Duffy, "Yellow Fever in the Continental United States during the Nineteenth Century," *Bulletin of the New York Academy of Medicine* 44 (June 1968): 687–701; Jo Ann Carrigan, *The Saffron Scourge: A History of Yellow Fever in Louisiana, 1796–1905* (Lafayette: Center for Louisiana Studies, University of Southwestern Louisiana, 1994), 82; Jo Ann Carrigan, "Yankees versus Yellow Jack in New Orleans, 1862–1866," *Civil War History* 9 (September 1963): 248–61; M. Lovell to Thomas O. Moore, May 12, 1862, *The War of the Rebellion: A Compilation of the Official Records of the Union and Confederate Armies,* vol. 15, ser. 1, 733–34, hereafter referred to as *OR; Harper's Weekly,* 17 May 1862.

3. Benjamin F. Butler, "Some Experiences with Yellow Fever and Its Prevention," *North American Review* 147 (November 1888): 525–41; Carrigan, "Yankees versus Yellow Jack," 250–53; Benjamin Butler, *Private and Official Correspondence of Gen. Benjamin F. Butler during the Period of the Civil War* (Norwood, Mass.: Plimpton Press, 1917), 2:112–13.

4. Elisha Harris, M.D., "Hygenic Experience in New Orleans during the War: Illustrating the Importance of Efficient Sanitary Regulations," *Southern Medical and Surgical Journal* 21 (July 1866): 77–88 (quotation); Butler, *Private and Official Correspondence,* 342; *OR,* vol. 2, ser. 3, 634–35 (quotation); *New Orleans Daily Picayune,* June 6, 1862; *Harper's Weekly,* August 16, 1862, and January 17, 1863.

5. Butler's policies did not eliminate all of the mosquito's potential breeding sites. Nineteenth-century New Orleanians kept their water in above-ground cisterns (wells are impractical in an area that rests below sea level), which may have also contained *Aedes aegypti* eggs. Margaret Humphreys, *Intensely Human: The Health of the Black Soldier in the American Civil War* (Baltimore: Johns Hopkins University Press, 2008), 104–5. Robert S. Holyman, "Ben Butler in the Civil War," *New England Quarterly* 30 (September 1957): 330–45; Michael A. Ross, "Justice Miller's Reconstruction: The Slaughter-House Cases, Health Codes, and Civil Rights in New Orleans, 1861–1873," *Journal of Southern History* 64 (November 1998): 649–76; *New Orleans Daily Picayune,* June 28, 1862.

6. L. Pierio to Andrew J. Hamilton, December 9, 1862, in Salmon P. Chase, *Diary*

and *Correspondence of Salmon P. Chase* (Washington, D.C.: American Historical Association, 1903), 428 (quotation); *Brownsville Flag* article reprinted in *Houston Tri-Weekly Telegraph*, November 12, 1862; *Brownsville Flag* editorial reprinted in *Galveston News*, October 8, 1862 (quotation); *Official Records of the Union and Confederate Navies in the War of the Rebellion*, vol. 19, ser. 1, 289–91, hereafter referred to as *ORN*; John Carrier, "Medicine in Texas: The Struggle with Yellow Fever, 1839–1903," *Texas Medicine* 82 (November 1986): 62–65; *Bellville Countryman*, October 18 and 25 and November 1, 1862; *Galveston Weekly News*, October 22, 1862; D. G. Farragut to Gideon Welles, October 9, 1862 (quotation), H. French to D. G. Farragut, September 18, 1862 (quotation), D. G. Farragut to H. French, October 7, 1862 (quotation), *ORN* vol. 19, ser. 1, 265, 289–93.

7. D. G. Farragut to W. B. Renshaw, September 19, 1862 (quotation), W. B. Renshaw to D. G. Farragut, October 8, 1862 (quotation), *ORN*, vol. 19, ser. 1, 213, 255–60; Charles C. Cumberland, "The Confederate Loss and Recapture of Galveston, 1862–1863," *Southwestern Historical Quarterly* 51 (October 1947): 109–19; Peggy H. Gregory, comp., *Record of Interments of the City of Galveston, 1859–1872* (Houston: privately printed, 1976), 28–29, in Rosenberg Library, Galveston; David G. McComb, *Galveston: A History* (Austin: University of Texas Press, 1986), 74–76.

8. *Houston Tri-Weekly Telegraph*, October 1 and 3 and November 7 and 10, 1862 (quotation); *Galveston Weekly News*, September 17, 1862; *ORN*, vol. 19, ser. 1, 260–61; H. P. Bee to Samuel Boyer Davis, November 15, 1862 (quotation), *OR*, vol. 15, ser. 1, 181–83; Kevin R. Young, *To the Tyrants Never Yield* (Plano, Tex.: Wordware Publishing, 1992), 123–27; *Austin State Gazette*, November 12, 1862.

9. *Houston Tri-Weekly Telegraph*, October 3 and 27 and November 3 and 7, 1862; W. T. Block, "Yellow Fever Plagued Area during the 1860s," *Beaumont Enterprise*, August 7, 1999; *Galveston News*, September 17 and 24, 1862 (quotation); *Natchitoches Union*, September 18, 1862; Andrew Forest Muir, "Dick Dowling and the Battle of Sabine Pass," in *Lonestar Blue and Gray: Essays on Texas in the Civil War*, ed. Ralph A. Wooster (Austin: Texas State Historical Association, 1995), 190; *Bellville Countryman*, September 6, 1862 (quotation); "Autobiography of Mrs. Otis McGaffey, Sr.," *Yellowed Pages* 28 (Fall 1998): 1–14 (quotation); Edward T. Cotham Jr., *Sabine Pass: The Confederacy's Thermopylae* (Austin: University of Texas Press, 2004), 29; Alwyn Barr, "Texas Coastal Defense, 1861–1865," in Wooster, *Lonestar*, 161; *OR*, vol. 15, ser. 1, 144–45, 813–15.

10. Lewis W. Pennington to William B. Renshaw, September 29, 1862, *ORN*, vol. 19, ser. 1, 219–24; *Galveston Weekly News*, October 15, 1862 (quotation); X. B. Debray to P. O. Hébert, September 28, 1862, *OR*, vol. 15, ser. 1, 815–17; Rodman L. Underwood, *Waters of Discord: The Union Blockade of Texas during the Civil War* (London: McFarland & Co., 2003), 83.

11. "Autobiography of Mrs. Otis McGaffey, Sr.," 7–8 (quotation); Francis Lubbock, *Six Decades in Texas or Memoirs of Francis Richard Lubbock, Governor of Texas in Wartime, 1861–63*, ed. C. W. Raines (Austin: Ben C. Jones & Co. Printers, 1900), 415; A. W. Spaight to R. M. Franklin, September 29 and October 2, 1862, *OR*, vol. 15, ser. 1, 145–47 (quotation); W. T. Block, "Yellow Fever Plagued Area during 1860s," *Beaumont Enterprise*, August 7, 1999.

It should be pointed out that in Spaight's official report of the battle, he argued that a larger Confederate force would not have made any difference at Sabine due to

Crocker's superior firepower. But with more men he might have at least temporarily resisted the landing. Instead, the U.S. Navy captured the city with no casualties.

12. It is possible that the small Houston outbreak of 1862 was ultimately caused by the larger epidemic that occurred in Sabine City the same year. The October 13, 1862, edition of the *Houston Tri-Weekly Telegraph* reported that a man and wife had died of yellow fever after returning from Beaumont, which received several evacuees from Sabine. *Houston Tri-Weekly Telegraph,* September 19 and October 22, 24, and 27, 1862; *Galveston Weekly News,* October 22 and 29, 1862; *San Antonio Semi-Weekly News,* October 23, 1862.

13. Lewis G. Schmidt, *The Civil War in Florida, A Military History,* vol. 3: *Florida's Keys & Fevers* (Allentown, Pa.: Lewis G. Schmidt, 1992), 254–55; G.R.B. Horner, M.D., fleet surgeon, "Notice of the Yellow Fever as It Occurred in Key West and in the U.S. East Gulf Blockading Squadron, in 1862," *American Journal of the Medical Sciences* 92 (October 1863): 391–98 (quotation); *ORN,* vol. 17, ser. 1, 294; *ORN,* vol. 1, ser. 1, 507–8.

14. *The Medical and Surgical History of the Civil War* (Wilmington, N.C.: Broadfoot Publishing Co., 1991), 6:676, hereafter referred to as *MSH;* Emily Holder, *At the Dry Tortugas during the Civil War,* quoted in Lewis G. Schmidt, *A Civil War History of the 47th Regiment of Pennsylvania Veteran Volunteers: The Wrong Place at the Wrong Time* (Allentown, Pa.: by the author, 1986), 146 (quotation); "The Situation," *New York Herald,* September 11, 1862; Horner, "Notice of the Yellow Fever," 391–98 (quotation); "Our Key West Correspondence.," *New York Times,* October 2 and November 12, 1862.

15. Schmidt, *History of the 47th Regiment,* 145–46; Book Records of Volunteer Union Organizations 90th New York Infantry, Companies A to C, Records of Adjutant General's Office, Record Group 94, National Archives and Records Administration, Washington, D.C.

16. *ORN,* vol. 19, ser. 1, 165, 184, 286–87 (quotation); *ORN,* vol. 27, ser. 1, 455; *ORN,* vol. 1, ser. 1, 493; Book Records of Union Volunteer Organizations 90th New York Infantry, Order Book, Companies A to F, Vol. 6, Record Group 94, National Archives; *MSH,* 5:679.

17. *MSH,* 2:235 (quotation); John N. Maffitt to "My Dear Daughter," September 8, 1862, and Maffitt to "My dear Florie," September 19, 1862, John Newland Maffitt Papers, University of North Carolina at Chapel Hill (quotation).

18. *MSH,* 5:678; Henry F. W. Little, *The Seventh Regiment, New Hampshire Volunteers in the War of the Rebellion* (Concord, N.H.: Ira C. Evans, Printer, 1896), 39–54 (quotation); Thomas T. Smiley, "The Yellow Fever at Port Royal, S.C.," *Boston Medical and Surgical Journal* 67, January 8, 1863, 449–68 (quotation); Frank R. Freemon, *Gangrene and Glory: Medical Care during the American Civil War* (Urbana: University of Illinois Press, 1998), 135–38.

19. Smiley, "Yellow Fever," 451–52; O. M. Mitchel to Halleck, September 20, 1862, *OR,* vol. 14, ser. 1, 383–84; F. A. Mitchel, *Ormsby MacKnight Mitchel, Astronomer and General: A Biographical Narrative* (Boston: Riverside Press, 1887), 358–82; McPherson, *Battle Cry,* 370–71; Russell McCormmach, "Ormsby MacKnight Mitchel's 'Sidereal Messenger,' 1846–1848," *Proceedings of the American Philosophical Society* 110, February 18, 1966, 35–47; Eicher, *Longest Night,* 234–35; Herman Hattaway and Archer Jones, *How the North Won: A Military History of the Civil War* (Urbana: University of Illinois Press, 1983), 206–7.

20. Smiley, "Yellow Fever," 452–54 (quotation); *MSH*, 5:678; John Bedel, "Historical Sketch of Third New Hampshire Volunteers," *Granite Monthly* 3 (September 1880): 513–34; Peter H. Buckingham, *All's for the Best: The Civil War Reminiscences and Letters of Daniel W. Sawtelle, Eighth Maine Volunteer Infantry* (Knoxville: University of Tennessee Press, 2001), 206; Herbert Beecher, *History of the First Light Battery, Connecticut Volunteers, 1861–1865* (New York: A. T. DeLaMare Ptg. and Pub. Co., 1905), 1:189 (quotation).

21. Willie Lee Rose, *Rehearsal for Reconstruction: The Port Royal Experiment* (London: Oxford University Press, 1964), xvi; Charlotte Forten, "Life on the Sea Islands," *Atlantic Monthly* 13 (May 1864): 587–96 (quotation); Charlotte Forten, *The Journal of Charlotte L. Forten,* with an introduction and notes by Ray Allen Billington (New York: Dryden Press, 1953), microfiche, 197 (quotation).

22. Rose, *Rehearsal,* 171–72 (quotation); Laura Towne, *Letters and Diary of Laura M. Towne. Written from the Sea Islands of South Carolina, 1862–1884,* ed. Rupert Sargent Holland (New York: Negro Universities Press, 1969), 39, 89 (quotation).

23. Elizabeth Ware Pearson, ed., *Letters from Port Royal Written at the Time of the Civil War* (Boston: W. B. Clarke, 1906), microfiche, 68, 73, 105 (quotation); Rose, *Rehearsal,* 171–72; Towne, *Letters,* 89 (quotation).

24. Smiley, "Yellow Fever," 460 (quotation); Mitchel, *Astronomer,* 379–82; "Death of General Mitchell," *Harper's Weekly,* November 15, 1862, 723; Isaiah Price, *History of the Ninety-seventh Regiment, Pennsylvania Volunteer Infantry, during the War of the Rebellion, 1861–65, with Biographical Sketches* (Philadelphia: by the author for the subscribers, 1875), 134–43 (quotation); *OR,* vol. 14, ser. 1, 388–89; Stephen Walkley, *History of the Seventh Connecticut Volunteer Infantry, Hawley's Brigade, Terry's Division, Tenth Army Corps, 1861–1865* (Hartford, 1905), 65.

25. Precisely when the *Kate* arrived and whether or not she was quarantined are subjects for debate since the sources provide conflicting accounts. See the *Wilmington Daily Journal,* September 26, 1862, W. T. Wragg, "Report on the Epidemic of Yellow Fever Which Prevailed at Wilmington, N.C., in the Fall of 1862," *New York Medical Journal* 9 (August 1869): 478–96; and an undated account written by Mrs. Charles P. Bolles, Charles P. Bolles Papers, Archives and Records Division, North Carolina Department of Cultural Resources, Raleigh (quotation); letter to A. J. Turlington, September 13, 1862, A. J. Turlington Papers, Duke University, Durham, N.C. (quotation).

26. *Wilmington Journal,* September 13 (quotation) and 27 and November 20, 1862; letters from William, October 23 (quotation) and October 28, 1862, De Rosset Family Papers, University of North Carolina, Chapel Hill; Todd L. Savitt, "Slave Health and Southern Distinctiveness," in *Disease and Distinctiveness in the American South,* ed. Todd L. Savitt and James Harvey Young (Knoxville: University of Tennessee Press, 1988), 123.

27. H. Drane to Mary Lindsay Hargrave Foxhall, October 23, 1862, Foxhall Family Papers, Southern Historical Collection, University of North Carolina, Chapel Hill; *Wilmington Journal,* October 2, 1862 (quotation).

28. Wragg, "Report," 478–85 (quotation); William Lofton to mother, October 20, 1862, William F. Lofton Papers, Duke University, Durham, N.C.; E. A. Anderson, "Yellow Fever as It Has Occurred in Wilmington, North Carolina, from 1800 to 1872, Being an Examination of the Review of Dr. Thomas, on the Report of Dr. Wragg," *New York*

Medical Journal 16 (September 1872): 225–59; Wilmington Journal, October 4, 1862; D. MacRae to "Dear Dix," October 1, 1862 (quotation), D. MacRae to Colonel John McRae, October 15, 1862, MacRae Family Papers, Duke University, Durham, N.C.

29. See "Siege Matters—Four Hundred and Seventy-fifth Day," Charleston Mercury, October 26, 1864.

30. Richard Everett Wood, "Port Town at War: Wilmington, North Carolina, 1860–1865" (Ph.D. diss., Florida State University, 1976), 176; W.H.C. Whiting to S. Cooper, January 15, 1863, and Whiting to George W. Randolph, November 14, 1862, OR, vol. 18, ser. 1, 848–49 (quotation), 774–76 (quotation); Wm. A. Parker to J. G. Foster, December 8, 1862, ORN, vol. 8, ser. 1, 263–64.

31. Wilmington Journal, November 27, 1862; Van Bokkelen to Donald MacRae, November 12, 1862, MacRae Family Papers (quotation); I. S. Murphy to Mary Ann Murphy, November 26, 1862, Cronly Family Papers, Duke University, Durham, N.C.; Wood, "Wilmington," 176; Patterson, "Yellow Fever Epidemics," 855–65; MSH, 6:679.

CHAPTER 4: "THE LAND OF FLOWERS, MAGNOLIAS, AND CHILLS"

1. T. H. Walker, "Camp Diarrhœa," Chicago Medical Journal 5 (August 1862): 478–80 (quotation); E. Andrews, "Army Correspondence," Chicago Medical Examiner 3 (August 1862): 479–84; Peter Josyph, ed., The Wounded River: The Civil War Letters of John Vance Lauderdale, M.D. (East Lansing: Michigan State University Press, 1993), 74–77 (quotation); Dr. J. S. Newberry, The U.S. Sanitary Commission in the Valley of the Mississippi, during the War of the Rebellion, 1861–1866 (Cleveland: Fairbanks, Benedict, & Co., 1871), 36–94 (quotation); William W. Belknap, History of the Fifteenth Regiment, Iowa Veteran Volunteer Infantry (Keokuk: R. B. Ogden & Son, 1887), 85.

2. Sherman's brief description of his bout with "malaria" does not make it clear that he was infected with plasmodium parasites. Because many of his men were infected, however, it is entirely possible that he was too. The Medical and Surgical History of the Civil War (Wilmington, N.C.: Broadfoot Publishing Co., 1991), 5:104, hereafter referred to as MSH; Rachel Sherman Thorndike, ed., The Sherman Letters: Correspondence between General Sherman and Senator Sherman from 1837 to 1891 (New York: Da Capo Press, 1969), 148; Paul E. Steiner, Disease in the Civil War: Natural Biological Warfare in 1861–1865 (Springfield, Ill.: Charles C. Thomas, 1968), 171–72; William T. Sherman, Memoirs of Gen. W. T. Sherman, Written by Himself (New York: Charles L. Webster & Co., 1891), 1:284; Edward Batwell, "Notes on the Fever That Prevailed amongst the Troops in Camp Big Springs, near Corinth, Miss., in June, 1862," Medical and Surgical Reporter 13 (December 1865): 364–65 (quotation).

3. The War of the Rebellion: A Compilation of the Official Records of the Union and Confederate Armies, vol. 10, ser. 1, pt. 1, 774–77, 784, hereafter referred to as OR; H. H. Cunningham, Doctors in Gray: The Confederate Medical Service (Baton Rouge: Louisiana State University Press, 1958), 181–82; Kate Cumming, A Journal of Hospital Life in the Confederate Army of Tennessee from the Battle of Shiloh to the End of the War (Louisville, Ky.: John P. Morton & Co., 1866), 18–22 (quotation); MSH, 5:104.

4. OR, vol. 16, ser. 1, pt. 2, 62–63 (quotation).

5. MSH, 5:169, 331; Benjamin F. Butler to H. W. Halleck, October 24, 1862, OR, vol.

15, ser. 1, 158–60; *United States Sanitary Commission Records, Series 7: Statistical Bureau Archives, Camp Inspection Returns, 1861–1864*, Manuscript Division, Library of Congress; *MSH*, 1:96–97.

6. Smith had been reinforced since May by Confederate John C. Breckenridge's divisions from Corinth; Earl Van Dorn assumed overall command. *OR*, vol. 15, ser. 1, 15–19; Bern Anderson, "The Naval Strategy of the Civil War," *Military Affairs* 26 (Spring 1962): 11–21; James McPherson, *Battle Cry of Freedom: The Civil War Era* (New York: Ballantine Books, 1988), 420–21; Shelby Foote, *The Civil War, a Narrative: Fort Sumter to Perryville* (New York: Vintage Books, 1986), 547–48.

7. *OR*, vol. 15, ser. 1, 25–26; Charles S. Foltz, *Surgeon of the Seas: The Adventurous Life of Surgeon General Jonathan M. Foltz in the Days of Wooden Ships* (Indianapolis: Bobbs-Merrill, 1931), 243–46; Steiner, *Disease*, 189; William C. Holbrook, *A Narrative of the Services of the Officers and Enlisted Men of the 7th Regiment of Vermont Volunteers* (New York: American Bank Note Co., 1882), 21–28 (quotation); Caroline E. Whitcomb, *History of the Second Massachusetts Battery (Nim's Battery) of Light Artillery, 1861–1865* (Concord, N.H.: Rumford Press, 1912), 33; Thomas Hamilton Murray, *History of the Ninth Regiment, Connecticut Volunteer Infantry* (New Haven: Price, Lee & Adkins Co., 1903), 111.

8. *MSH*, 1:153–54; Holbrook, *Vermont*, 22–29; Murray, *Connecticut*, 109; *Harper's Weekly*, August 2, 1862; *OR*, vol. 15, ser. 1, 31–33.

9. T. Williams to C. H. Davis, July 24, 1862, *Extract from Diary of Flag-Officer Davis*, July 31, 1862, *ORN*, vol. 23, ser. 1, 239–41, 270–72 (quotation).

10. E. Kirby Smith to Braxton Bragg, July 24, 1862, *OR*, vol. 16, ser. 1, pt. 2, 734–35 (quotation). The following year Smith informed his subordinate in the Trans-Mississippi Department, Major General John Bankhead Magruder, that "the yellow fever" and "malaria of Lower Louisiana" would pressure the Union forces at New Orleans into launching a campaign up the Red River Valley "with an eye to the establishment of bases of operations against Texas." Union general Nathaniel Banks led just such a campaign in the spring of 1864 but mainly because of political arm-twisting and pressure from northern manufacturers who wanted access to Black Belt cotton. See E. Kirby Smith to J. B. Magruder, June 11, 1863, *OR*, vol. 26, ser. 1, pt. 2, 47–48.

11. Leon F. Litwack, *Been in the Storm So Long: The Aftermath of Slavery* (New York: Vintage Books, 1979), 66–68; E. Andrews, "Army Correspondence," *Chicago Medical Examiner* 3 (August 1862): 479–84 (quotation);

12. Joseph T. Glatthaar, *Forged in Battle: The Civil War Alliance of Black Soldiers and White Officers* (New York: Free Press, 1990), 7–8; Benjamin Butler, *Private and Official Correspondence of Gen. Benjamin F. Butler during the Period of the Civil War* (Norwood, Mass.: Plimpton Press, 1917), 2:143–44 (quotation)

13. *OR*, vol. 15, ser. 1, 34–39; Salmon P. Chase to Butler, July 31, 1862, in Butler, *Private and Official Correspondence*, 2:131–32 (quotation); G. Mott Williams, "Letters of General Thomas Williams, 1862," *American Historical Review* 14 (January 1909): 304–28 (Williams quoted p. 325).

14. Van Dorn's confidence stemmed in part from the success of the CSS *Arkansas*, a Confederate ironclad that had triumphantly engaged Farragut's fleet during the Vicksburg campaign. Earl Van Dorn to District of Mississippi Headquarters, September 9, 1862, *OR*, vol. 15, ser. 1, 15–19 (quotation), 76–81, 1125–26; Ed. Porter Thompson, *History of the First Kentucky Brigade* (Cincinnati: Caxton Publishing House, 1868), 119,

158; Steiner, *Disease*, 188; Dr. W. J. Worsham, *The Old Nineteenth Tennessee Regiment, C.S.A., June 1861. April, 1865.* (Knoxville: Press of Paragon Printing Co., 1902), 57–59 (quotation); Mary Elizabeth Sanders, ed. "Letters of a Confederate Soldier, 1862–1863," *Louisiana Historical Quarterly* 29 (October 1946): 1229–40.

15. *OR*, vol. 15, ser. 1, 14–19, 76–81; Steiner, *Disease*, 203; Major John B. Pirtle, "Defence of Vicksburg in 1862—The Battle of Baton Rouge," in *Southern Historical Society Papers*, vol. 8: *January to December, 1880* (Richmond, Va.: Rev. J. Williams Jones, D.D.), 327–28; Worsham, *Tennessee*, 59; G. Mott Williams, "The First Vicksburg Expedition and the Battle of Baton Rouge, 1862." in *War Papers Read before the Commandery of the State of Wisconsin, Military Order of the Loyal Legion of the United States* (Milwaukee: Burdick, Armitage & Allen, 1896), 57–61; Holbrook, *Vermont*, 50; Foote, *Sumter*, 579; Ira B. Gardner, "Personal Experiences with the Fourteenth Maine Volunteers from 1861–1865," in *War Papers Read before the Commandery of the State of Maine, Military Order of the Loyal Legion of the United States* (Portland, Maine: Lefavor-Tower Co., 1915), 4:90–113.

16. Foote, *Sumter*, 580–82; Butler, *General Orders No. 57*, August 9, 1862, *OR*, vol. 15, ser. 1, 15–19, 41–42 (quotation), 76–81.

17. Again, precise diagnoses are impossible without a lab test. *MSH*, 5:331; G. W. Randolph to R. E. Lee, August 4, 1862, *OR*, vol. 15, ser. 1, 794 (quotation).

18. David J. Eicher, *The Longest Night: A Military History of the Civil War* (New York: Simon & Schuster, 2001), 256–57; Price, *Pennsylvania*, 127; Herbert Beecher, *History of the First Light Battery, Connecticut Volunteers, 1861–1865* (New York: A. T. DeLaMare Ptg. And Pub. Co., 1905), 1:157–70 (quotation); William Todd, *The Seventy-ninth Highlanders: New York Volunteers in the War of the Rebellion, 1861–1865* (Albany, N.Y.: Brandow, Barton, & Co., 1886), 144–66; D. Hunter to H. G. Wright, June 27, 1862, *OR*, vol. 14, ser. 1, 43–47 (quotation).

19. Samuel Logan, "Prophylactic Effects of Quinine," *Confederate States Medical & Surgical Journal* 1 (June 1864): 81–83; Todd, *Highlanders*, 144 (quotation); Beecher, *Connecticut*, 169; "The Wee Nee Volunteers of Williamsburg District, South Carolina, in the First (Hagood's) Regiment, by Major John G. Pressley, of the Eutaw Battalion, South Carolina Volunteers," in *Southern Historical Society Papers* (Richmond, Va.: Southern Historical Society, 1888), 16:126–52; "Extracts from the Diary of Lieutenant-Colonel John G. Pressley, of the Twenty-fifth South Carolina Volunteers," in *Southern Historical Society Papers* (Richmond: Rev. J. William Jones, D.D., 1886), 14:35–62; T. A. Washington to J. C. Pemberton, April 2, 1862, *OR*, vol. 6, ser. 1, 423–24.

20. W.H.C. Whiting to Seddon, May 30, 1863, *OR*, vol. 18, ser. 1, 1081; Mark F. Boyd, "An Historical Sketch of the Prevalence of Malaria in North America," *American Journal of Tropical Medicine* 21 (March 1941): 223–44 (quotation); George Worthington Adams, *Doctors in Blue: The Medical History of the Union Army in the Civil War* (Baton Rouge: Louisiana State University Press, 1980), 243; D. W. Hand, *Extracts from Reports Relative to the Operations of the Medical Staff in the Department of North Carolina, from August, 1863, to the Close of the War*, in *MSH*, 2:238–41(quotation); "Medical History of the Seventeenth Regiment Mass. Volunteers," *Boston Medical and Surgical Journal* 68 (February 1863): 136–41 (quotation); Thomas H. Parker, *History of the 51st Regiment of P.V. and V.V.* (Philadelphia: King & Baird, Printers, 1869), 120–25 (quotation); Stanley B. Weld, "A Connecticut Surgeon in the Civil War: The Reminiscences of Dr. Nathan Mayer," *Journal of the History of Medicine and Allied Sciences* 19 (July 1964): 272–86 (quotation).

21. The importance of McClellan's defeat during the Peninsular Campaign cannot be overstated and is best described by the country's preeminent scholar on the Civil War era, James M. McPherson, who believes "this Confederate success convinced Lincoln to 'take off the kid gloves' in dealing with slavery and to adopt emancipation as a means of weakening the Confederacy and strengthening the Union cause." James M. McPherson, *Crossroads of Freedom: Antietam* (New York: Oxford University Press, 2002), xvi; *OR*, vol. 14, ser. 1, 528–49; *OR*, vol. 28, ser. 1, pt. 2, 444–45; *OR*, vol. 53, ser. 1, 300; Dan L. Morrill, *The Civil War in the Carolinas* (Charleston: Nautical & Aviation Publishing Co. of America, 2002), 187.

22. Joseph K. Barnes, *The Medical and Surgical History of the War of the Rebellion, 1861–1865. Part First* (Washington, D.C.: Government Printing Office, 1870), 55, 192; Steiner, *Disease,* 75; Robert E. Denney, *Civil War Medicine: Care & Comfort of the Wounded* (New York: Sterling Publishing, 1994), 174; John H. Dirckx, ed., *Stedman's Concise Medical Dictionary for the Health Professions,* 4th ed. (Philadelphia: Lippincott, Williams, & Wilkins, 2001), 159; Adams, *Doctors in Blue,* 201.

23. Cunningham, *Doctors in Gray,* 191; Joseph I. Waring, *A History of Medicine in South Carolina, 1825–1900* (Charleston: South Carolina Medical Association, 1967), 127–39; Logan, "Prophylactic," 83; *MSH,* 5:104–5.

24. O. M. Mitchel to H. W. Halleck, October 24, 1862, *OR*, vol. 14, ser. 1, 144–47; F. A. Mitchel, *Ormsby MacKnight Mitchel, Astronomer and General: A Biographical Narrative* (Boston: Riverside Press, 1887), 370; Eicher, *Longest,* 235; Richard E. Beringer, Herman Hattaway, Archer Jones, and William N. Still Jr., *Why the South Lost the Civil War* (Athens: University of Georgia Press, 1986), 199–200.

CHAPTER 5: "THE PESTILENT MARSHES OF THE PENINSULA"

1. Darrett B. Rutman and Anita H. Rutman, "Of Agues and Fevers: Malaria in the Early Chesapeake," *William and Mary Quarterly* 33 (January 1976): 31–60; Wyndham B. Blanton, *Medicine in Virginia in the Nineteenth Century* (Richmond, Va.: Garrett & Massie, 1933), 258–59; J. Bankhead Magruder to S. Cooper, August 21, 1861, *The War of the Rebellion: A Compilation of the Official Records of the Union and Confederate Armies,* vol. 51, ser. 1, pt. 2, 246, hereafter referred to as *OR*.

2. Regis De Trobriand, *Four Years with the Army of the Potomac* (Boston: Ticknor & Co., 1889), 221–36 (quotation); J. Theodore Calhoun, "The Chickahominy and Seven Pines," *Medical and Surgical Reporter* 9 (March 1863): 399–400 (quotation); Francis J. Parker, *The Story of the Thirty-second Regiment Massachusetts Infantry* (Boston: C. W. Calkins & Co., 1880), 51–57 (quotation); William Child, *A History of the Fifth Regiment New Hampshire Volunteers* (Bristol: R. W. Musgrove Printer, 1893), 60–96 (quotation).

3. Thomas P. Lowry, ed., *Swamp Doctor: The Diary of a Union Surgeon in the Virginia and North Carolina Marshes* (Mechanicsburg, Pa.: Stackpole Books, 2001), 226 (quotation); George Worthington Adams, *Doctors in Blue: The Medical History of the Union Army in the Civil War* (Baton Rouge: Louisiana State University Press, 1980), 71; *OR*, vol. 11, ser. 1, pt. 1, 201–2; Calhoun, "Chickahominy," 399–400; Paul E. Steiner, *Disease in the Civil War: Natural Biological Warfare in 1861–1865* (Springfield, Ill.: Charles C. Thomas, 1968), 108; G. W. Randolph to A. T. Bledsoe, August 26, 1861, *OR*, 51, ser. 1, pt. 2, 251.

4. Dale C. Smith, "The Rise and Fall of Typhomalarial Fever: I. Origins," *Journal of*

the *History of Medicine and Allied Sciences* 37 (April 1982): 182–220; Frank R. Freemon,
Gangrene and Glory: Medical Care during the American Civil War (Chicago: University of
Illinois Press, 1998), 70; Steiner, *Disease,* 124–25; Alfred Jay Bollet, *Civil War Medicine:
Challenges and Triumphs* (: Galen Press, 2002), 276–77.

5. *OR,* vol. 11, ser. 1, pt. 1, 82–84, 181, 210–20; Steiner, *Disease,* 124; Jonathan
Letterman, *Medical Recollections of the Army of the Potomac* (1866; rpt., Bohemian Bri-
gade Publishers, 1994), 6–14; James McPherson, *Battle Cry of Freedom: The Civil War
Era* (New York: Ballantine Books, 1988), 525; *OR,* vol. 11, ser. 1, pt. 3, 313–14, 331–33;
George Meade, *The Life and Letters of George Gordon Meade, Major-General United States
Army* (New York: Charles Scribner's Sons, 1913), 302–3.

6. Many blacks developed a limited immunity to the local malarial parasites in their
neighborhoods through repeated infections and/or possessed "innate immunities based
on inherited hemoglobin variants." Nineteenth-century physicians noticed that blacks
were less likely to die from yellow fever. Modern historians disagree over the causes of
this immunity. Some believe people of African descent inherited a genetic resistance
to yellow fever, while others argue that blacks' frequent contact with the disease in
southern towns gave them acquired immunity. Margaret Humphreys, *Intensely Human:
The Health of the Black Soldier in the American Civil War* (Baltimore: Johns Hopkins University Press, 2008), 49–50, 84–85 (quotation).

7. *OR,* vol. 16, ser. 1, pt. 2, 62–63; McPherson, *Battle Cry,* 488–89.

8. H. H. Cunningham, *Doctors in Gray: The Confederate Medical Service* (Baton
Rouge: Louisiana State University Press, 1958), 157, 191; *Medical Department Register of
Patients, General Hospital No. 21. Richmond, Virginia, 1862. Chapter VI, Volume 253,* Record
Group 109, National Archives, Washington, D.C.; *Medical Department [Confederate] Reg-
ister of Patients, General Hospital No. 18 Richmond, Virginia, 1862–1863. Chapter VI, Volume
217½,* Record Group 109, National Archives, Washington, D.C.; Herbert M. Nash, M.D.,
"Some Reminiscences of a Confederate Surgeon," *Transactions of the College of Physicians
of Philadelphia* 28 (Philadelphia: printed for the College, 1906): 122–44 (quotation).

9. See Bruce Catton, *Grant Moves South* (Boston: Little, Brown, 1960); R. E. Lee
to James A. Seddon, May 10, 1863, *OR,* vol. 25, ser. 1, pt. 2, 790; *Chicago Times* article
reprinted in *Wilmington Journal,* September 4, 1862.

10. McPherson, *Battle Cry,* 517; Frank R. Freemon, M.D., "The Medical Challenge of
Military Operations in the Mississippi Valley during the American Civil War," *Military
Medicine* 157 (September 1992): 494–97; *OR,* vol. 24, ser. 1, pt. 3, 357–58; Report of
Surgeon Madison Mills, June 20, 1863, *United States Sanitary Commission Records,* ser. 7:
Statistical Bureau Archives, Camp Inspection Returns, 1861–64, Manuscript Division,
Library of Congress; Captain Joseph Edward King, "Shoulder Straps for Aesculapius:
The Vicksburg Campaign, 1863," *Military Surgeon* 114 (March 1954), 216–26; T. H.
Barton, *Autobiography of Dr. Thomas H. Barton, the Self-Made Physician of Syracuse, Ohio,
Including a History of the Fourth Regt. West Va. Vol. Infy.* (Charleston: West Virginia Print-
ing Co., 1890), 151.

11. For more information on Grant's attempt to finish Williams's canal, see David
F. Bastian, *Grant's Canal: The Union Attempt to Bypass Vicksburg* (Shippensburg, Pa.:
Burd Street Press, 1995); Freemon, "Medical Challenge," 495; Frank R. Freemon, M.D.,
"Medical Care at the Siege of Vicksburg, 1863," *Bulletin of the New York Academy of
Medicine* 67 (September–October 1991): 429–38; H. S. Hewitt to Madison Mills, March

16, 1863, *United States Sanitary Commission Records,* ser. 7: Statistical Bureau Archives, Camp Inspection Returns, 1861–64, Manuscript Division, Library of Congress, Washington, D.C. (quotation); Granville P. Conn, *History of the New Hampshire Surgeons in the War of the Rebellion* (Concord, N.H.: Ira C. Evans Co., Printers, 1906), 209 (quotation); Lewis Crater, *History of the Fiftieth Regiment, Penna. Vet. Vols., 1861–65.* (Reading, Pa.: Coleman Printing House, 1884), 42 (quotation); A. J. Withrow to Libertatia Withrow, June 3, 1863, Withrow Papers, University of North Carolina, Chapel Hill (quotation); Charles F. Larimer, ed., *Love and Valor: Intimate Civil War Letters between Captain Jacob and Emeline Ritner* (Western Springs, Ill.: Sigourney Press, 2000), 173–74 (quotation); Thomas H. Parker, *History of the 51st Regiment of P.V. and V.V.* (Philadelphia: King & Baird Printers, 1869), 322.

12. Freemon, "Medical Care," 432; George C. Osborn, ed., "A Tennessean at the Siege of Vicksburg: The Diary of Samuel Alexander Ramsey Swan, May–July, 1863," *Tennessee Historical Quarterly* 14 (December 1955): 353–72; Rev. Albert Theodore Goodloe, *Confederate Echoes: A Voice from the South in the Days of Secession and of the Southern Confederacy* (Nashville: Smith & Lamar, 1907), 248–49; Kenneth Trist Urquhart, ed., *Vicksburg, Southern City under Siege: William Lovelace Foster's Letter Describing the Defense and Surrender of the Confederate Fortress on the Mississippi* (New Orleans: Historic New Orleans Collection, 1980), 35.

13. Some historians have suggested that Lee was either ignorant of the conditions in the Deep South or using the "sickly season" as an excuse to swell the size of the army protecting his native Virginia. But this theory ignores his prewar experience as an army engineer in coastal Georgia and the fact that many other Civil War commanders, both Union and Confederate, shared his assessment of the southern climate. See Richard M. McMurry, "Marse Robert and the Fevers: A Note on the General as Strategist and on Medical Ideas as a Factor in Civil War Decision Making," *Civil War History* 35 (September 1989): 197–207; Lee to Seddon, June 8, 1863, *OR,* vol. 27, ser. 1, pt. 3, 868–69; Seddon to W. Porcher Miles, May 13, 1863, *OR,* vol. 14, ser. 1, 940; Edward Younger, *Inside the Confederate Government: The Diary of Robert Garlick Hill Kean, Head of the Bureau of War* (New York: Oxford University Press, 1957), 63; Beauregard to Seddon, May 11, 1863, *OR,* vol. 14, ser. 1, 933–35.

14. Leon F. Litwack, *Been in the Storm So Long: The Aftermath of Slavery* (New York: Vintage Books, 1979), 67–70; Martha M. Bigelow, "The Significance of Milliken's Bend in the Civil War," *Journal of Negro History* 45 (July 1960): 156–63 (quotation); *OR,* vol. 24, ser. 1, pt. 3, 156–57 (quotation); "The Week," *American Medical Times* 7, August 8, 1863, 65–66 (quotation); William A. Hammond, *A Treatise on Hygiene with Special Reference to the Military Service* (1863; rpt., San Francisco: Norman Publishing, 1991), ix–x, 70 (quotation); Willie Lee Rose, *Rehearsal for Reconstruction: The Port Royal Experiment* (London: Oxford University Press, 1964), xvi; Humphreys, *Intensely Human,* 45, 105.

15. *The Medical and Surgical History of the Civil War* (Wilmington, N.C.: Broadfoot Publishing Co., 1991), 5:85 (quotations), hereafter referred to as *MSH.*

16. S. M. Barton to C. L. Stevenson, July 1, 1863, *OR,* vol. 24, ser. 1, pt. 2, 347.

17. A. Cumming to C. L. Stevenson & A. W. Reynolds to C. L. Stevenson, July 1, 1863, *OR,* vol. 24, ser. 1, pt. 2, 348–49; Freemon, "Medical Care," 434; John C. Pemberton, *Pemberton: Defender of Vicksburg* (Chapel Hill: University of North Carolina Press, 1942), 222–24; Judith Lee Hallock, "'Lethal and Debilitating': The Southern Disease

Environment as a Factor in the Confederate Defeat," *Journal of Confederate History* 7 (1991): 51–61.

18. Freemon, *Gangrene*, 225; Freemon, "Medical Challenge," 495;William T. Sherman, General Orders, No. 69, August 30, 1863, *OR*, vol. 30, ser. 1, pt. 3, 225–26 (quotation); Report of Medical Inspector E.O.F. Roler, September 27, 1863, *United States Sanitary Commission Records*, ser. 7: Statistical Bureau Archives, Camp Inspection Returns, 1861–64, Manuscript Division, Library of Congress, Washington, D.C.

19. J. H. Clarke, "Yellow Fever in New Orleans," *Boston Medical and Surgical Journal* 59, November 12, 1863, 303–4; H. H. Bell to N. P. Banks, October 23, 1863, *Official Records of the Union and Confederate Navies in the War of the Rebellion*, vol. 20, ser. 1, 638, 650–52 (quotation), hereafter referred to as *ORN; ORN*, vol. 25, ser. 1, 461–62; Jo Ann Carrigan, "Yankees versus Yellow Jack in New Orleans, 1862–1866," *Civil War History* 9 (September 1963): 256.

20. Clarke, "Yellow Fever," 303; *ORN* 20, ser. 1, 603; *Hospital Tickets and Case Papers, Naval Hospital, New Orleans, 1863*, Record Group 52, National Archives (quotation); *Naval Hospital New Orleans, La., Hospital Tickets and Case Papers, 1863*, Record Group 52, National Archives, Washington, D.C.

21. B. F. Gibbs, "Account of the Epidemic of Yellow Fever Which Visited Pensacola Navy Yard in the Summer and Autumn of 1863," *American Journal of the Medical Sciences* 61 (1866): 340–51; George F. Pearce, *Pensacola during the Civil War* (Gainesville: University Press of Florida, 2000), 193; William C. Holbrook, *A Narrative of the Services of the Officers and Enlisted Men of the 7th Regiment of Vermont Volunteers* (New York: American Bank Note Co., 1882), 134–35 (quotation).

22. H. H. Bell to John P. Gillis, September 11, 1863, *ORN*, vol. 20, ser. 1, 509–10, 566–67 (quotation), 581–83, 637–38, 708; *ORN*, vol. 21, ser. 1, 240–43, 284, 328, 679, 692; *ORN*, vol. 27, ser. 1, 595–96; Holbrook, *Narrative*, 134–35 (quotation); letter to Louis James M. Boyd, November 5, 1863, Louis James M. Boyd Papers, State Archives of Florida, Tallahassee (quotation); Jo Ann Carrigan, "Yankees versus Yellow Jack in New Orleans, 1862–1866," Civil War History 9 (September 1963): 256; Elisha Harris, M.D., "Hygiene Experience in New Orleans during the War: Illustrating the Importance of Efficient Sanitary Regulations," *Southern Medical and Surgical Journal* 21 (July 1866): 86; Jo Ann Carrigan, *The Saffron Scourge: A History of Yellow Fever in Louisiana, 1796–1905* (Lafayette: Center for Louisiana Studies, University of Southwestern Louisiana, 1994), 82.

23. See Allan Nevins, *The War for the Union: The Improvised War, 1861–1862* (New York: Charles Scribner's Sons, 1959); and *The War for the Union: The Organized War, 1863–1864* (New York: Charles Scribner's Sons, 1971); Adams, *Doctors*, 31–34; Bollett, *Medicine*, 236–37.

CHAPTER 6: "THE ROUGHEST TIMES ANY SET OF SOLDIERS EVER ENCOUNTERED"

1. Reid Mitchell, *The American Civil War, 1861–1865* (Harlow, Eng.: Pearson Education, 2001), 56; Charles W. Johnson, "Narrative of the Sixth Regiment" (1891), in *Minnesota in the Civil and Indian Wars, 1861–1865* by Minnesota Board of Commissioners on Publication of History of Minnesota in Civil and Indian Wars (St. Paul, Minn.: Pioneer Press Co., 1890–93), 300–328 (quotation); Charles H. Lothrop, M.D., *A History of the*

First Regiment Iowa Cavalry Veteran Volunteers (Lyons, Iowa: Beers & Eaton, 1890), 130; *Illinois at Vicksburg* (published under the authority of an act of the Forty-fifth General Assembly by the Illinois-Vicksburg Military Park Commission, 1907), Library of Congress, Washington, D.C., 141–312; E. B. Quiner, *The Military History of Wisconsin* (Chicago: Clarke & Co., Publishers, 1866), 760–66; C. C. Andrews, "Narrative of the Third Regiment," in Johnson, *Minnesota*, 165–67; Frederick Steele to John Schofield, September 12, 1863, *The War of the Rebellion: A Compilation of the Official Records of the Union and Confederate Armies*, vol. 22, ser. 1, pt. 1, 474–77, hereafter referred to as *OR* (quotation).

2. John Scott, *Story of the Thirty-second Iowa Infantry Volunteers* (Nevada, Iowa: John Scott, 1896), 59; Shelby Foote, *The Civil War: A Narrative* (New York: Vintage Books, 1986), 2:702; A. F. Sperry, *History of the 33d Iowa Infantry Volunteer Regiment, 1863–6* (Des Moines: Mills & Co., 1866), 37–41 (quotation).

3. Frederick Steele to Stephen A. Hurlbut, August 23, 1863, Steele to John Schofield, September 1, 1863, *OR*, vol. 22, ser. 1, pt. 1, 472–74; Leander Stillwell, *The Story of a Common Soldier of Army Life in the Civil War, 1861–1865*, 2nd ed. ([Erie, Kan.]: Franklin Hudson Publishing Co., 1920), 157–58 (quotation).

4. Joseph R. Smith to J. K. Barnes, January 20, 1866, *Sanitary Report of the Army of Arkansas for the Last Quarter of 1863*, U.S. Surgeon General's Office, History of Medicine Division, National Library of Medicine, Bethesda, Md. (quotation); *The Medical and Surgical History of the Civil War* (Wilmington, N.C.: Broadfoot Publishing Co., 1991), 2:343–46 (quotation), hereafter referred to as *MSH*.

5. Scott, *Iowa*, 59–61; Sperry, *33d Iowa*, 42–43 (quotation); Frederick Steele to H. W. Halleck, September 12, 1863, *OR*, vol. 22, ser. 1, pt. 1, 474–82 (quotation); Paul E. Steiner, *Disease in the Civil War: Natural Biological Warfare in 1861–1865* (Springfield, Ill.: Charles C. Thomas, 1968), 218–19.

6. H. W. Halleck to E.R.S. Canby, November 12, 1864, *OR*, vol. 41, ser. 1, pt. 4, 529 (quotation); F. Steele to J. M. Thayer, November 12, 1864, *OR*, vol. 41, ser. 1, pt. 4, 535; "Report of Troops Serving in the Department of Arkansas," October 31, 1864, *OR*, vol. 41, ser. 1, pt. 4, 341; Steiner, *Disease*, 219.

7. See Surgeon Eugene F. Sanger, "Extracts from the Report of the Medical Director of the Nineteenth Corps, for April, 1864," in *The Medical and Surgical History of the War of the Rebellion (1861–1865)* (Washington, D.C.: Government Printing Office, 1870), 335.

8. Confederate successes kept Steele from reinforcing General Nathaniel Banks, who also suffered a humiliating loss during the Red River Campaign of 1864. A. C. Wedge quoted in C. C. Andrews, "Narrative of the Third," in *Minnesota*, 174–75; Christopher C. Andrews, *Pioneer in Forestry Conservation in the United States: For Sixty Years a Dominant Influence in the Public Affairs of Minnesota: Lawyer: Editor: Diplomat: General in the Civil War. Recollections: 1829–1922*, ed. Alice E. Andrews (Cleveland: Arthur H. Clark Co., 1928), 191 (quotation); Stephen Miller to E.R.S. Canby, April 4, 1865, *OR*, vol. 48, ser. 1, pt. 2, 31–32.

9. *OR*, vol. 41, ser. 1, pt. 2, 711–16; W. P. Belden to Stephen Miller, October 29, 1864, reprinted in Johnson, "Narrative of the Sixth Regiment," 322–26 (quotation); *OR*, vol. 41, ser. 1, pt. 3, 502; Arthur M. Daniels, *A Journal of Sibley's Indian Expedition during the Summer of 1863 and Record of the Troops Employed by a Soldier in Company "H," Sixth Regiment* (Minneapolis: James D. Thueson Publisher, 1980), 151; Alfred J. Hill, *History of Company E of the Sixth Minnesota Regiment of Volunteer Infantry* (1899; rpt., St. Paul,

Minn.: Budd Parrish, 1992), 28 (quotation); Peter H. Wood, *Black Majority: Negroes in Colonial South Carolina from 1670 through the Stono Rebellion* (New York: W. W. Norton, 1974), 72–75, 86; *OR*, vol. 41, ser. 1, pt. 1, 190–91.

10. Dr. Benjamin Woodward to Dr. Elisha Harris, September 15, 1865, *United States Sanitary Commission Records, Series 1: Medical Committee Archives, 1861–1866*, Library of Congress, Washington, D.C.; H. W. Halleck to Reynolds, January 26, 1865, *OR*, vol. 48, ser. 1, pt. 1, 649.

11. Joseph Palmer Blessington, *The Campaigns of Walker's Texas Division* (Austin: State House Press, 1994), 42 (quotation); Junius N. Bragg to "My Dear Josephine," December (?), 1862, and Bragg to "My Darling Wife," August 18, 1863, in J. N. Bragg, *Letters of a Confederate Surgeon, 1861–65* (Camden, Ark.: Hurley Co., 1960), 101–4, 169 (quotation); William M. McPheeters, *I Acted from Principle: The Civil War Diary of Dr. William M. McPheeters, Confederate Surgeon in the Trans-Mississippi*, ed. Cynthia Dehaven Pitcock and Bill J. Gurley (Fayetteville: University of Arkansas Press, 2002), 41–63, 94–95 (quotation); *OR*, vol. 22, ser. 1, pt. 1, 439.

12. J. Bankhead Magruder to J. F. Belton, May 7, 1863, *OR*, vol. 15, ser. 1, 1078–79 (quotation).

13. *Austin State Gazette*, September 28, October 12 and 26, November 23, and December 7, 14, and 24, 1864; *Houston Tri-Weekly Telegraph*, November 16, 1864; "The Diary of H. C. Medford, Confederate Soldier, 1864," *Southwestern Historical Quarterly* 34 (1930): 106–40; Peggy H. Gregory, comp., *Record of Interments of the City of Galveston, 1859–1872* (Houston: privately printed, 1976), 34, Rosenberg Library, Galveston, Tex.; Charles W. Hayes, *History of the Island and the City of Galveston* (Austin, Tex.: Jenkins Garrett Press, 1974), 2:625–26; Charles C. Cumberland, "The Confederate Loss and Recapture of Galveston, 1862–1863," *Southwestern Historical Quarterly* 51 (October 1947): 118; Alwyn Barr, "Texas Coastal Defense, 1861–1865," in *Lonestar Blue and Gray: Essays on Texas in the Civil War*, ed. Ralph A. Wooster (Austin: Texas State Historical Association, 1995), 14–15; David G. McComb, *Galveston: A History* (Austin: University of Texas Press, 1986), 79; H. C. Medord, "The Diary of H. C. Medford, Confederate Soldier, 1864," *Southwestern Historical Quarterly* 34 (1930): 106–40; *Austin State Gazette*, September 21 (quotation) & October 12, 1864; Charles W. Hayes, *History of the Island and the City of Galveston* (Austin, Tex.: Jenkins Garrett Press, 1974), 2:623–25; *OR*, vol. 41, ser. 1, pt. 2, 1020; D. G. Farragut to W. E. Le Roy, September 21, 1864, *Official Records of the Union and Confederate Navies in the War of the Rebellion*, vol. 21, ser. 1, 655, hereafter referred to as *ORN*.

14. *Austin State Gazette*, December 7, 14, and 24, 1864; Galveston Interment Records, 43–49; J. G. Walker to W. R. Boggs, October 26, 1864, *OR*, vol. 41, ser. 1, pt. 4, 1014–15 (quotation); Hayes, *Galveston*, 649–51; K. David Patterson, "Yellow Fever Epidemics and Mortality in the United States, 1693–1905," *Social Science and Medicine* 34 (1992): 858.

15. *ORN*, vol. 27, ser. 1, 595–96; Jo Ann Carrigan, "Yankees versus Yellow Jack in New Orleans, 1862–1866," *Civil War History* 9 (September 1963): 256; Elisha Harris, M.D., "Hygenic Experience in New Orleans during the War: Illustrating the Importance of Efficient Sanitary Regulations," *Southern Medical and Surgical Journal* 21 (July 1866): 86; Jo Ann Carrigan, *The Saffron Scourge: A History of Yellow Fever in Louisiana, 1796–1905* (Lafayette: Center for Louisiana Studies, University of Southwestern Louisiana, 1994), 82; *U.S. Navy, 1775–1910, Special Reports, Epidemics, Special Incidents, Etc. 1830–1910*, Record Group 45, National Archives, Washington, D.C.

16. Fort Jefferson continued to experience yellow fever epidemics after the war was over. Dr. Samuel Mudd, the physician who famously set John Wilkes Booth's leg after the Lincoln assassination, was imprisoned in the fort and aided victims during an 1867 outbreak. Emily Holder, *At the Dry Tortugas during the War*, quoted in Lewis G. Schmidt, *The Civil War in Florida: A Military History* (Allentown, Pa.: by the author, 1992), 3:868–69 (quotation); "The Situation," *New York Herald*, November 2, 1864.

17. *New York Times*, August 31, 1864 (quotation); John A. Wilder to mother, June 19, 1864, and Wilder to brother-in-law, August 2, 1864, in Millicent Todd Bingham, ed., "Key West in the Summer of 1864," *Florida Historical Quarterly* 43 (1965): 262–65 (quotation); Book Records of Union Volunteer Organizations, Second USCT Infantry Regimental Order, Letter, Endorsement, Deaths, Discharges, and Miscellaneous Book, Record Group 94, National Archives, Washington, D.C.; Book Records of Volunteer Union Organizations, Second USCT Infantry Descriptive Book, Records of the Adjutant General's Office, Record Group 94, National Archives, Washington, D.C.

18. *ORN*, vol. 21, ser. 1, 277, 601, 614–15: *ORN*, vol. 27, ser. 1, 625, 629; Theodorus Bailey to Gideon Welles, July 27, 1864, Welles to Bailey, July 19, 1864, *ORN*, vol. 17, ser. 1, 723, 733, 737 (quotation); Schmidt, *Civil War in Florida*, 860; Edward S. Miller, "A Long War and a Sickly Season," *Blue & Gray Magazine* 9 (June 1992): 40–44.

19. James O. Breeden, "A Medical History of the Later Stages of the Atlanta Campaign," *Journal of Southern History* 35 (February 1969): 31–59; C. S. Frink, "Extracts from a Report on the Operations of the Medical Department of the Third Division of the Twenty-third Corps from June 11, to September 10, 1864," in *The Medical and Surgical History of the War of the Rebellion (1861–1865)* (Washington, D.C.: Government Printing Office, 1870), 318 (quotation); Jasper E. James, *Letters from a Civil War Soldier*, ed. Vera Dockery Elkins (New York: Vantage Press, 1969), 52 (quotation); Stephen Davis, *Atlanta Will Fall: Sherman, Joe Johnston, and the Yankee Heavy Battalions* (Wilmington, Del.: Scholarly Resources, 2001), 33–34.

20. *MSH*, 5:105; W.A.W. Spotswood to S. R. Mallory, November 30, 1863, *ORN*, vol. 2, ser. 2, 559–61; Dabney H. Maury to S. Cooper, November 10, 1864, *OR*, vol. 39, ser. 1, pt. 3, 910–11 (quotation); James McPherson, *Battle Cry of Freedom: The Civil War Era* (New York: Ballantine Books, 1988), 761; Confederate Medical Department, Register of Patients, Ross Hospital, Mobile, Ala., 1863–65, Record Group 109, National Archives, Washington, D.C.; Foote, *Civil War*, 3:967.

21. Alfred Jay Bollet, M.D., *Plagues & Poxes: The Impact of Human History on Epidemic Disease* (New York: Demos, 2004), 37; Margaret Humphreys, *Malaria: Poverty, Race, and Public Health in the United States* (Baltimore: Johns Hopkins University Press, 2001), 11–13; Steiner, *Disease*, 219.

CHAPTER 7: BIOLOGICAL WARFARE

1. *New York Times*, May 30, 1865 (quotation); William A. Tidwell, James O. Hall, and David Winfred Gaddy, *Come Retribution: The Confederate Secret Service and the Assassination of Lincoln* (Jackson: University Press of Mississippi, 1988), 186–87.

2. Nancy Disher Baird, *Luke Pryor Blackburn: Physician, Governor, Reformer* (Lexington: University Press of Kentucky, 1979), Tidwell, *Retribution*, 185; Edward Steers Jr., "Risking the Wrath of God," *North & South* 3 (September 2000): 59–70.

3. Edward Steers Jr., "A Rebel Plot and Germ Warfare," *Washington Times*, November 10, 2001; Ben. Perley Poore, ed., *The Conspiracy Trial for the Murder of the President, and the Attempt to Overthrow the Government by the Assassination of Its Principal Officers* (Boston: J. E. Tilton and Co., 1865), 2:409–19 (quotation).

4. Blackburn was not the only Confederate interested in using yellow fever against northerners. Early in the war a Louisianan named R. R. Barrow advocated sending blankets infected with yellow fever into New Orleans in order to trigger an epidemic that would end the Union occupation there. See R. R. Barrow to D. F. Kenner, September 11, 1862, John T. Pickett Papers, Manuscript Division, Library of Congress, Washington, D.C.; "The Soiled Clothes Investigation in St. Georges," *Bermuda Royal Gazette*, April 25, 1865 (quotation); Steers, "Wrath," 64–67; John W. Headley, *Confederate Operations in Canada and New York* (New York: Neale Publishing Co., 1906), 264–65; Tidwell, *Come Retribution*, 186; Baird, *Blackburn*, 8; K. J. Stewart to Jefferson Davis, November 30 and December 12, 1864, *Prison Pens, Canada Raids, Secret Operations, 1864*, Record Group 109, chap. 7, vol. 24, National Archives, Washington, D.C.

5. Thomas J. Farnham and Francis P. King, "'The March of the Destroyer': The New Bern Yellow Fever Epidemic of 1864," *North Carolina Historical Review* 73 (October 1996): 435–83; Stanley B. Weld, "A Connecticut Surgeon in the Civil War: The Reminiscences of Dr. Nathan Mayer," *Journal of the History of Medicine and Allied Sciences* 19 (July 1964): 284–85; Sheldon B. Thorpe, *The History of the Fifteenth Connecticut Volunteers in the War for the Defense of the Union, 1861–1865* (New Haven, Conn.: Price, Lee & Adkins, 1893), 233; *The Medical and Surgical History of the Civil War* (Wilmington, N.C.: Broadfoot Publishing Co., 1991), 5:679–682, hereafter referred to as *MSH;* James Gifford to Parents, September 26, 1864, James Gifford Papers, University of North Carolina at Chapel Hill.

6. Alan D. Watson, *A History of New Bern and Craven County* (New Bern: Tryon Palace Commission, 1987), 408–9; Thomas Kirwan, Memorial *History of the Seventeenth Regiment Massachusetts Volunteer Infantry (Old and New Organizations) in the Civil War from 1861–1865* (Salem, Mass.: Salem Press Co., 1911), 247–51; Thorpe, *History of the Fifteenth Connecticut Volunteers*, 234–36; *MSH*, 2:239; Farnham and King, "March," 456–60; Judkin Browning and Michael Thomas Smith, eds., *Letters from a North Carolina Unionist: John A. Hedrick to Benjamin S. Hedrick, 1862–1865* (Raleigh: North Carolina Division of Archives and History, 2001), 227–28; *MSH*, 5:679; W. S. Benjamin, *The Great Epidemic in New Berne and Vicinity, September and October, 1864, by One Who Passed through It* (New Bern, N.C.: Geo. Mills Joy, 1865), 16–19.

7. *Official Records of the Union and Confederate Navies in the War of the Rebellion*, vol. 8, ser. 1, 263–64, hereafter referred to as *ORN; The War of the Rebellion: A Compilation of the Official Records of the Union and Confederate Armies*, vol. 35, ser. 1, pt. 2, 310–11, hereafter referred to as *OR;* Lonnie R. Speer, *Portals to Hell: Military Prisons of the Civil War* (Mechanicsburg, Pa.: Stackpole Books, 1997), 215; *Charleston Daily Courier*, October 5, 11, 18, and 20, 1864; *Charleston Mercury*, November 7 and 8, 1864; *OR*, vol. 42, ser. 1, pt. 3, 1151–53; Grant to General Ruger, July 7, 1865, *OR*, vol. 47, ser. 1, pt. 3, 675–76.

8. Farnham and King, "March," 435–483; Benjamin, *Great Epidemic*, 10–11; "The Yellow Fever Plot: Newbern, N.C., May 31, 1865," *Bermuda Royal Gazette*, June 13, 1865 (quotation).

9. Morphine (pain reliever), ether, and chloroform (anesthetics) were also in high

demand across the South. Michael A. Flannery, *Civil War Pharmacy: A History of Drugs, Drug Supply and Provision, and Therapeutics for the Union and Confederacy* (New York: Pharmaceutical Products Press, 2004), 193; William Diamond, "Imports of the Confederate Government from Europe and Mexico," *Journal of Southern History* 6 (November 1940): 470–503; Mary Elizabeth Massey, *Ersatz in the Confederacy: Shortages and Substitutes on the Southern Homefront* (Columbia: University of South Carolina Press, 1993), 11, 120; Norman H. Franke, "Rx Prices in the Confederacy," *Journal of the American Pharmaceutical Association,* no. 12 (December 1961): 773–74; Joseph Jacobs, "Some of the Drug Conditions during the War between the States, 1861–5," *Southern Historical Papers* 33 (Richmond, Va.: Southern Historical Society, 1905), 175.

10. Francis B. Simkins and James W. Patton, "The Work of Southern Women among the Sick and Wounded of the Confederate Armies," *Journal of Southern History* 1 (November 1935): 475–96; Louise Llewellyn Jarecka, "Virginia Moon, Unreconstructed Rebel," *Delphian Quarterly* 30 (January 1947): 17–21 (quotation); Elizabeth D. Leonard, *All the Daring of the Soldier: Women of the Civil War Armies* (New York: W. W. Norton, 1999), 72, 82 (quotation); Mary Elizabeth Massey, *Bonnet Brigades* (New York: Alfred A. Knopf, 1966), 105–6; Alan Axelrod, *The War between the Spies: A History of Espionage during the American Civil War* (New York: Atlantic Monthly Press, 1992), 32; J. Hampton Hoch, "Through the Blockade," *Journal of the American Pharmaceutical Association* 12 (December 1961): 769–70.

11. *OR,* vol. 2, ser. 2, 329 (quotation); *OR,* vol. 4, ser. 2, 880–85 (quotation); Willard Carver, *Fourteenth Regt., Maine Infantry. Roster of Survivors with Abstract of Regimental History* (1890), 2 (quotation), Library of Congress, Washington, D.C.; Joseph H. Parks, "A Confederate Trade Center under Federal Occupation: Memphis, 1862 to 1865," *Journal of Southern History* 7 (August 1941): 289–314.

12. Yellow fever occasionally disrupted the activities of these runners (and the Federal ships that hunted them) because they operated out of ports where the disease was endemic. The 1864 outbreak in Bermuda, for example, prompted the U.S. consul there to declare the island's "blockade-running business . . . to be at an end for the season." See *ORN,* vol. 3, ser. 1, 161–63 and 447.

13. Frank E. Vandiver, *Confederate Blockade Running through Bermuda, 1861–1865: Letters and Cargo Manifests* (Austin: University of Texas Press, 1947), 109; Thomas E. Taylor, *Running the Blockade: A Personal Narrative of Adventures, Risks, and Escapes during the American Civil War* (Annapolis, Md.: Naval Institute Press, 1995), xxvi, 18, 65; *Wilmington Daily Journal,* October 1, 1863; Flannery, *Pharmacy,* 200 (quotation); Massey, *Ersatz,* 115.

14. *OR,* vol. 1, ser. 4, 1041; H. H. Cunningham, *Doctors in Gray: The Confederates Medical Service* (Baton Rouge: Louisiana State University Press, 1958), 29, 146–48; Francis Peyre Porcher, *Resources of the Southern Fields and Forests* (1863; rpt., New York: Arno Press, 1970), 38–411; *OR,* vol. 2, ser. 4, 1024; George Worthington Adams, "Confederate Medicine," *Journal of Southern History* 6 (May 1940), 151–66; *Charleston Daily Courier,* August 11, 1862; Massey, *Ersatz,* 120.

15. Flannery, *Pharmacy,* 227; *New Orleans Daily Picayune,* August 4, 1861; Guy R. Hasegawa, "Pharmacy in the American Civil War," *American Journal of Health-System Pharmacy* 57 (March 2000): 475–89; Anna DeWolf Middleton to Anna E. Marston DeWolf, September 19, 1864, Nathaniel Russell Middleton Papers, in *Records of Ante-bellum*

microfilm (Frederick, Md.: University Publications of America, 1985–2000), ser. J, pt. 3.

16. *OR,* vol. 2, ser. 4, 1024; *OR,* vol. 35, ser. 1, pt. 2, 592–93; Cunningham, *Gray,* 157, 191; *Harper's Weekly,* August 27, 1864, 547; David J. Eicher, *The Longest Night: A Military History of the Civil War* (New York: Simon & Schuster, 2001), 723; Herbert M. Nash, M.D., "Some Reminiscences of a Confederate Surgeon," *Transactions of the College of Physicians of Philadelphia* 28 (Philadelphia: printed for the College, 1906), 135–36; *ORN,* vol. 10, ser. 1, 350–53, 730–50; *OR,* vol. 42, ser. 1, pt. 2, 1271.

17. These percentages include cases of "remittent fever," "quotidian intermittent fever," "tertian intermittent fever," "quartan intermittent fever," and "congestive intermittent fever." "Other diseases of this [Miasmatic] Order" (447 cases) are not included. *MSH,* vol. 1, 174–78, 490–91; *Harper's Weekly,* March 11, 1865, 149; George Worthington Adams, *Doctors in Blue: The Medical History of the Union Army in the Civil War* (Baton Rouge: Louisiana State University Press, 1980), 104; *OR,* vol. 17, ser. 1, pt. 2, 640.

18. Flannery, *Civil War Pharmacy,* 163–68; Paul E. Steiner, *Disease in the Civil War: Natural Biological Warfare in 1861–1865* (Springfield, Ill.: Charles C. Thomas, 1968), 22.

EPILOGUE

1. *OR,* vol. 14, ser. 1, 539; *OR,* vol. 25, ser. 1, pt. 2, 790. See also Alexander H. H. Stuart to George W. Randolph, April 28, 1862, *OR,* vol. 51, ser. 1, pt. 2, 550–51.

2. K. David Patterson, "Yellow Fever Epidemics and Mortality in the United States, 1693–1905," *Social Science and Medicine* 34 (1992): 855–65; Margaret Humphreys, *Yellow Fever and the South* (Baltimore: Johns Hopkins University Press, 1992), 60–64.

3. Scientists are only now beginning to understand the extent of genetic diversity within the *Plasmodium* genus. See "Genetic Map Offers New Tool for Malaria Research," *Science Daily,* December 12, 2006, www.sciencedaily.com/releases/2006/12/061211092750.htm (January 20, 2009).

4. John Duffy "The Impact of Malaria on the South," in *Disease and Distinctiveness in the American South,* ed. Todd L. Savitt and James Harvey Young (Knoxville: University of Tennessee Press, 1988), 29–54. For more information on New England's postwar epidemic, see "The Increase of Malaria," *Boston Daily Advertiser,* June 22, 1882; and "Malaria. By a Massachusetts Physician," *Boston Congregationalist,* January 14, 1886.

5. Leslie Poles Hartley, *The Go-Between* (London: H. Hamilton, 1953).

6. Christian F. Ockenhouse, Alan Magill, Dale Smith, and Wil Milhous, "History of U.S. Military Contributions to the Study of Malaria," *Military Medicine* 170 (April 2005): 12–16; Ashley M. Croft, Alicia H. Darbyshire, Christopher J. Jackson, and Pieter P. van Thiel, "Malaria Prevention Measures in Coalition Troops in Afghanistan," *Journal of the American Medical Association* 297, May 23–30, 2007, 2197–99; Aldo Leopold, *A Sand County Almanac* (New York: Oxford University Press, 1949).

SELECTED BIBLIOGRAPHY

PRIMARY SOURCES

Manuscripts

Charles P. Bolles Papers, North Carolina Department of Cultural Resources
Louis James M. Boyd Papers, State Archives of Florida, Tallahassee
Cronly Family Papers, Duke University, Durham, N.C.
De Rosset Family Papers, University of North Carolina at Chapel Hill
Foxhall Family Papers, University of North Carolina at Chapel Hill
James Gifford Papers, University of North Carolina at Chapel Hill
William F. Lofton Papers, Duke University, Durham, N.C.
MacRae Family Papers, Duke University, Durham, N.C.
John Newland Mafitt Papers, University of North Carolina at Chapel Hill
John T. Pickett Papers, Library of Congress
Zachary Taylor Papers, 1820–50, Library of Congress
A. J. Turlington Papers, Duke University, Durham, N.C.
Adoniram Judson Withrow Papers, University of North Carolina at Chapel Hill

Published Sources

Alexander, Edward Porter. *Fighting for the Confederacy: The Personal Recollec-*
tions of General Edward Porter Alexander. Chapel Hill: University of North
Carolina Press, 1989.

Anderson, E. A. "Yellow Fever as It Has Occurred in Wilmington, North
Carolina, from 1800 to 1872, Being an Examination of the Review of Dr.
Thomas, on the Report of Dr. Wragg." *New York Medical Journal* 16 (Sep-
tember 1872): 225–59.

Anderson, Thomas. *Handbook for Yellow Fever: Describing Its Pathology and*
Treatment. London: Churchill & Sons, 1856.

Andrews, Christopher C. *Pioneer in Forestry Conservation in the United States:*
For Sixty Years a Dominant Influence in the Public Affairs of Minnesota:
Lawyer: Editor: Diplomat: General in the Civil War. Recollections: 1829–1922.
Edited by Alice E. Andrews. Cleveland: Arthur H. Clark Co., 1928.

Andrews, E. "Army Correspondence." *Chicago Medical Examiner* 3 (August 1862): 479–84.

"The Annual Report of the Board of Health, for the Year 1850, as Required by Law." *New Orleans Medical and Surgical Journal* 8 (March 1851): 590–91.

Armstrong, George D. *The Summer of Pestilence: A History of the Ravages of the Yellow Fever in Norfolk, Virginia, A.D. 1855*. Philadelphia: J. B. Lippincott & Co., 1856.

Arnold, Richard D. *An Essay upon the Relation of Bilious and Yellow Fever, Prepared at the Request of, and Read before the Medical Society of the State of Georgia, at Its Session Held at Macon*. Augusta, Ga.: J. Morris, 1856.

Bartlett, Elisha. *The History, Diagnosis, and Treatment of the Fevers of the United States*. Philadelphia: Blanchard & Lea, 1852.

Barton, T. H. *Autobiography of Dr. Thomas H. Barton, The Self-Made Physician of Syracuse, Ohio, Including a History of the Fourth Regt. West Va. Vol. Infy*. Charleston: West Virginia Printing Co., 1890.

Batwell, Edward. "Notes on the Fever That Prevailed amongst the Troops in Camp Big Springs, Near Corinth, Miss., in June, 1862." *Medical and Surgical Reporter* 13 (December 1865): 364–65.

Bedel, John. "Historical Sketch of Third New Hampshire Volunteers." *Granite Monthly* 3 (September 1880): 513–34.

Beecher, Herbert. *History of the First Light Battery, Connecticut Volunteers, 1861–1865*, Vol. 1. New York: A. T. DeLaMare Ptg. and Pub. Co., 1905.

Belknap, William W. *History of the Fifteenth Regiment, Iowa Veteran Volunteer Infantry*. Keokuk: R. B. Ogden & Son, 1887.

Benjamin, W. S. *The Great Epidemic in New Berne and Vicinity, September and October, 1864, by One Who Passed through It*. New Bern, N.C.: Geo. Mills Joy, 1865.

Bingham, Millicent Todd, ed. "Key West in the Summer of 1864." *Florida Historical Quarterly* 43 (1965): 262–65.

Black, J. R. "On the Ultimate Cause of Malarial Disease." *St. Louis Medical and Surgical Journal* 12 (July 1854): 353–63.

Bragg, J. N. *Letters of a Confederate Surgeon, 1861–65*. Camden, Ark.: Hurley Co., 1960.

Brinton, John. *Personal Memoirs of John H. Brinton, Major and Surgeon U.S.V., 1861–1865*. New York: Neale Publishing Co., 1914.

Browning, Judkin, and Michael Thomas Smith, eds. *Letters from a North Carolina Unionist: John A. Hedrick to Benjamin S. Hedrick, 1862–1865*. Raleigh: North Carolina Division of Archives and History, 2001.

Bryan, James. "Negro Regiments—Department of the Tennessee." *Boston Medical & Surgical Journal* 69 (1864): 43–44.

Buckingham, Peter H. *All's for the Best: The Civil War Reminiscences and Letters of Daniel W. Sawtelle, Eighth Maine Volunteer Infantry.* Knoxville: University of Tennessee Press, 2001.

Butler, Benjamin. *Private and Official Correspondence of Gen. Benjamin F. Butler during the Period of the Civil War,* Vol. 2. Norwood, Mass.: Plimpton Press, 1917.

———. "Some Experiences with Yellow Fever and Its Prevention." *North American Review* 147 (November 1888): 525–41.

Cain, D. J. *History of the Epidemic of Yellow Fever in Charleston, S.C., in 1854.* Philadelphia: T. K. & P. G. Collins, 1856.

Calhoun, J. Theodore. "The Chickahominy and Seven Pines." *Medical and Surgical Reporter* 9 (March 1863): 399–400.

Carver, Willard. *Fourteenth Regt. Maine Infantry. Roster of Survivors with Abstract of Regimental History.* N.p.p., 1890.

Castleman, Alfred Lewis. *The Army of the Potomac.* Milwaukee: Strickland & Co., 1863.

Chase, Salmon P. *Diary and Correspondence of Salmon P. Chase.* Washington, D.C.: American Historical Association, 1903.

Chenery, William H. *The Fourteenth Regiment Rhode Island Heavy Artillery (Colored,) In the War to Preserve the Union, 1861–1865.* 1898. Reprint. New York: Negro Universities Press, 1969.

Child, William. *A History of the Fifth Regiment New Hampshire Volunteers.* Bristol: R. W. Musgrove Printer, 1893.

Clarke, J. H. "Yellow Fever in New Orleans." *Boston Medical and Surgical Journal* 59, November 12, 1863, 303–4.

Conn, Granville P. *History of the New Hampshire Surgeons in the War of the Rebellion.* Concord: Ira C. Evans Co., Printers, 1906.

Conway, Moncure D. *The True and the False in Prevalent Theories of Divine Dispensations.* Washington, D.C.: Taylor & Maury, 1855.

Coxe, Edward Jenner. *Practical Remarks on Yellow Fever, Having Special Reference to the Treatment.* New Orleans: J. C. Morgan, 1859.

Crater, Lewis. *History of the Fiftieth Regiment, Penna. Vet. Vols., 1861–65.* Reading, Pa.: Coleman Printing House, 1884.

Cumming, Kate. *A Journal of Hospital Life in the Confederate Army of Tennessee from the Battle of Shiloh to the End of the War.* Louisville, Ky.: John P. Morton & Co., 1866.

Daniels, Arthur M. *A Journal of Sibley's Indian Expedition during the Summer of 1863 and Record of the Troops Employed by a Soldier in Company "H," Sixth Regiment.* Minneapolis: James D. Thueson Publisher, 1980.

Davis, Varina. *Jefferson Davis, Ex-President of the Confederate States of America; A Memoir by His Wife.* New York: Belford Co., 1890.

DeRolette, Frs. Xavier. *Lecture on the Fever and Ague and Other Intermittent Fevers.* Pittsburgh: A. A. Anderson & Sons, 1865.

De Trobriand, Regis. *Four Years with the Army of the Potomac.* Boston: Ticknor & Co., 1889.

Dickens, Charles. *American Notes for General Circulation.* 1842. Reprint. London: Penguin Books, 2000.

Dodson, C. Marion. *Yellow Flag: The Civil War Journal of Surgeon's Steward C. Marion Dodson.* Edited by Charles Albert Earp. Baltimore: Maryland Historical Society, 2002.

Douglass, Frederick. *Narrative of the Life of Frederick Douglass, an American Slave, Written by Himself.* 1845. Reprint. New York: New American Library, 1997.

Dowdey, Clifford, and Louis H. Manarin, eds. *The Wartime Papers of Robert E. Lee.* New York: Da Capo Press, 1961.

Drake, Daniel. *A Systematic Treatise, Historical, Etiological, and Practical, on the Principal Diseases of the Interior Valley of North America, as They Appear in the Caucasian, African, Indian, and Esquimaux Varieties of Its Population.* 1850. Reprint. Chicago: University of Illinois Press, 1964.

Eve, Paul F. "Answers to Certain Questions Propounded by Prof. Charles A. Lee, M.D., Agent of the United States Sanitary Commission, relative to the Health, &c., of the late Southern Army." *Nashville Journal of Medicine and Surgery* 1 (July 1866): 12–32.

"The Fate of Virginia." *Living Age* 70, August 10, 1861, 374–75.

Felder, W. L. "Observations on the Yellow Fever Epidemic of 1854, in Augusta, Ga." *Southern Medical and Surgical Journal* 11 (October 1855): 598–608.

Fleming, Walter L., ed. *General W. T. Sherman as College President.* Cleveland: Arthur M. Clark Co., 1912.

Forten, Charlotte. *The Journal of Charlotte L. Forten.* Intro. and notes by Ray Allen Billington. New York: Dryden Press, 1953.

———. "Life on the Sea Islands." *Atlantic Monthly* 13 (May 1864): 587–96.

Gaillard, Edwin Samuel. *An Essay on Intermittent and Bilious Remittent Fevers: With Their Pathological Relation to Ozone.* Charleston: Walker & Evans, 1856.

Gibbs, B. F. "Account of the Epidemic of Yellow Fever Which Visited Pensacola Navy Yard in the Summer and Autumn of 1863." *American Journal of the Medical Sciences* 51 (1866): 340–51.

Goodloe, Albert Theodore. *Confederate Echoes: A Voice from the South in the Days of Secession and of the Southern Confederacy.* Nashville: Smith & Lamar, 1907.

Goyne, Minetta Altgelt. *Lone Star and Double Eagle: Civil War Letters of a German-Texas Family.* Fort Worth: Texas Christian University Press, 1982.

Gregory, Peggy H., comp. *Record of Interments of the City of Galveston, 1859–1872.* Houston: Privately Printed, 1976. Housed at Rosenberg Library, Galveston.

Harris, Elisha. "Hygenic Experience in New Orleans during the War: Illustrating the Importance of Efficient Sanitary Regulations." *Southern Medical and Surgical Journal* 21 (July 1866): 77–88.

Headley, John W. *Confederate Operations in Canada and New York.* New York: Neale Publishing Co., 1906.

Higginson, Thomas Wentworth. *Army Life in a Black Regiment and Other Writings.* New York: Penguin Books, 1997.

Hill, Alfred J. *History of the Company E of the Sixth Minnesota Regiment of Volunteer Infantry.* 1899. Reprint. St. Paul, Minn.: Budd Parrish, 1992.

Holbrook, William C. *A Narrative of the Services of the Officers and Enlisted Men of the 7th Regiment of Vermont Volunteers.* New York: American Bank Note Co., 1882.

Holton, William C. *Cruise of the U.S. Flag-ship Hartford, 1862–1863.* 1863. Reprint. Tarrytown, N.Y.: W. Abbatt, 1922.

Horner, G.R.B. "Notice of the Yellow Fever as It Occurred in Key West and in the U.S. East Gulf Blockading Squadron, in 1862." *American Journal of the Medical Sciences* 92 (October 1863): 391–98.

Howe, Henry Warren. *Passages from the Life of Henry Warren Howe, Consisting of Diary and Letters Written during the Civil War, 1816–1865.* Lowell, Mass.: Courier-Citizen Co., Printers, 1899.

Illinois at Vicksburg: Published under the Authority of an Act of the Forty-fifth General Assembly by the Illinois-Vicksburg Military Park Commission. Chicago: Blakely Printing Co., 1907.

James, Jasper E. *Letters from a Civil War Soldier.* Edited by Vera Dockery Elkins. New York: Vantage Press, 1969.

Johnson, Charles W. "Narrative of the Sixth Regiment" (1891). In *Minnesota in the Civil and Indian Wars, 1861–1865* by Minnesota Board of Commissioners on Publication of History of Minnesota in Civil and Indian Wars. St. Paul: Pioneer Press Co., 1890–93.

Johnson, Jonathan Huntington. *The Letters and Diary of Captain Jonathan Huntington Johnson.* N.p.: Alden Chase Brett, 1961.

Jones, C. Handfield. "Considerations Respecting the Operation of Malaria on the Human Body." *Association Medical Journal* 187 (August 1856): 668–70.

Josyph, Peter, ed. *The Wounded River: The Civil War Letters of John Vance Lauderdale, M.D.* East Lansing: Michigan State University Press, 1993.

Kirwin, Thomas. *Memorial History of the Seventeenth Regiment Massachusetts Volunteer Infantry (Old and New Organizations) in the Civil War from 1861–1865.* Salem, Mass.: Salem Press Co., 1911.

Larimer, Charles F., ed. *Love and Valor: Intimate Civil War Letters between Captain Jacob and Emeline Ritner.* Western Springs, Ill.: Sigourney Press, 2000.

LaRoche, Rene. *Pneumonia: Its Supposed Connection, Pathological and Etiologi-*

cal, with Autumnal Fevers; Including an Inquiry into the Existence and Morbid Agency of Malaria. Philadelphia: Blanchard & Lea, 1854.

Leavenworth, F. P. "Malaria." *St. Louis Medical and Surgical Journal* 12 (January 1854): 35–58.

Letterman, Jonathan. *Medical Recollections of the Army of the Potomac.* 1866. Reprint. Knoxville: Bohemian Brigade Publishers, 1994.

Logan, Samuel. "Prophylactic Effects of Quinine." *Confederate States Medical and Surgical Journal* 1 (June 1864): 81–83.

Lothrop, Charles H. *A History of the First Regiment Iowa Cavalry Veteran Volunteers.* Lyons, Iowa: Beers & Eaton, 1890.

Lowry, Thomas P., ed. *Swamp Doctor: The Diary of a Union Surgeon in the Virginia and North Carolina Marshes.* Mechanicsburg, Pa.: Stackpole Books, 2001.

Lubbock, Francis. *Six Decades in Texas or Memoirs of Francis Richard Lubbock, Governor of Texas in War-time, 1861–63.* Edited by C. W. Raines. Austin, Tex.: Ben C. Jones & Co., Printers, 1900.

Martin, W. H. "Various Forms of Intermittent Diseases." *Illinois and Indiana Medical and Surgical Journal* 2 (1847–48): 407–9.

McCraven, William. "The Epidemic of Yellow Fever of 1859, in Houston, Texas." *New Orleans Medical News and Hospital Gazette* 7 (April 1860): 105–10.

McGaffey, Mrs. Otis, Sr. "Autobiography of Mrs. Otis McGaffey, Sr." *Yellowed Pages* 28 (Fall 1998): 1–14.

McGregor, Charles. *History of the Fifteenth Regiment, New Hampshire Volunteers, 1862–1863.* Published by Order of the Fifteenth Regiment Association. [Concord, N.H.: I. C. Evans], 1900.

McPheeters, William M. *I Acted from Principle: The Civil War Diary of Dr. William M. McPheeters, Confederate Surgeon in the Trans-Mississippi.* Edited by Cynthia Dehaven Pitcock and Bill J. Gurley. Fayetteville: University of Arkansas Press, 2002.

Meade, George. *The Life and Letters of George Gordon Meade, Major-General United States Army.* New York: Charles Scribner's Sons, 1913.

Medford, H. C. "The Diary of H. C. Medford, Confederate Soldier, 1864." *Southwestern Historical Quarterly* 34 (1930): 106–40.

The Medical and Surgical History of the Civil War. 15 vols. Wilmington, N.C.: Broadfoot Publishing Co., 1991.

"Medical History of the Seventeenth Regiment Mass. Volunteers." *Boston Medical and Surgical Journal* 68 (February 1863): 136–41.

Merrill, A. P. *Lectures on Fever, Delivered in the Memphis Medical College, in 1853–6.* New York: Harper & Bros., 1865.

Monroe, Haskell M., Jr., and James T. McIntosh. *The Papers of Jefferson Davis,* vol. 1: *1808–1840.* Baton Rouge: Louisiana State University Press, 1971.

Murray, Thomas Hamilton. *History of the Ninth Regiment, Connecticut Volunteer Infantry.* New Haven: Price, Lee, & Adkins Co., 1903.

Nance, Hiram. "Retrospect of Miasmatic Diseases for a Few Years Past." *Northwestern Medical and Surgical Journal* 3 (February 1854): 54–57.

Nash, Herbert M. "Some Reminiscences of a Confederate Surgeon." *Transactions of the College of Physicians of Philadelphia* 28 (1906).

Naval Records Collection of the Office of Naval Records and Library. Record Group 45, National Archives. Washington, D.C.

Newberry, J. S. *The U.S. Sanitary Commission in the Valley of the Mississippi, during the War of the Rebellion, 1861–1866.* Cleveland: Fairbanks, Benedict, & Co., 1871.

Official Records of the Union and Confederate Navies in the War of the Rebellion. 30 vols. Washington, D.C.: Government Printing Office, 1894–1922.

Olmstead, Frederick Law. *The Cotton Kingdom.* 1861. Reprint. New York: Random House, 1984.

"On the External Application of Oil of Turpentine as a Substitute for Quinine in Intermittent Fever, with Reports of Cases." *Confederate States Medical and Surgical Journal* 1 (January 1864): 7–8.

Osborn, George C., ed. "A Tennessean at the Siege of Vicksburg: The Diary of Samuel Alexander Ramsey Swan, May–July, 1863." *Tennessee Historical Quarterly* 14 (December 1955): 353–72.

Parker, Francis J. *The Story of the Thirty-second Regiment Massachusetts Infantry.* Boston: C. W. Calkins & Co., 1880.

Parker, Thomas H. *History of the 51st Regiment of P.V. and V.V.* Philadelphia: King & Baird, Printers, 1869.

Pearson, Elizabeth Ware, ed. *Letters from Port Royal Written at the Time of the Civil War.* Boston: W. B. Clarke, 1906.

Pemberton, John C. *Pemberton: Defender of Vicksburg.* Chapel Hill: University of North Carolina Press, 1942.

Pepper, William. *On the Use of Bebeerine and Cinchonia in the Treatment of Intermittent Fever.* Philadelphia: T. K. and P. G. Collins, 1853.

Poore, Ben. Perley, ed. *The Conspiracy Trial for the Murder of the President, and the Attempt to Overthrow the Government by the Assassination of Its Principal Officers,* Vol. 2. Boston: J. E. Tilton and Co., 1865.

Porcher, Francis Peyre. *Resources of the Southern Fields and Forests.* 1863. Reprint. New York: Arno Press, 1970.

Price, Isaiah. *History of the Ninety-seventh Regiment, Pennsylvania Volunteer Infantry, during the War of the Rebellion, 1861–65, with Biographical Sketches.* Philadelphia: By the Author for the Subscribers, 1875.

Quaife, Milo Milton, ed. *Growing Up with Southern Illinois 1820 to 1861 from the Memoirs of Daniel Harmon Brush.* Chicago: Lakeside Press, 1944.

Quiner, E. B. *The Military History of Wisconsin.* Chicago: Clarke & Co., Publishers, 1866.

Records of the Adjutant General's Office. Record Group 94, National Archives. Washington, D.C.

Records of the Bureau of Medicine and Surgery. Record Group 52, National Archives. Washington. D.C.

Records of the Office of the Surgeon General (Army). Record Group 112, National Archives. Washington. D.C.

Report of the Philadelphia Relief Committee Appointed to Collect Funds for the Sufferers by Yellow Fever, at Norfolk and Portsmouth, Va., 1855. Philadelphia: Inquirer Printing Office, 1856.

Report of the Portsmouth Relief Association to the Contributors of the Fund for the Relief of Portsmouth, Virginia, during the Prevalence of Yellow Fever in That Town in 1855. Richmond: H. K. Ellyson's Steam Power Presses, 1856.

Report on the Origin of Yellow Fever in Norfolk during the Summer of 1855. Made to the City Councils by a Committee of Physicians. Richmond, Va.: Ritchie and Dunnavant, 1857.

Robinson, William L. *The Diary of a Samaritan. By a Member of the Howard Association of New Orleans.* New York: Harper & Bros., 1860.

Rogers, Stephen. *Quinine as a Prophylactic or Protective from Miasmatic Poisoning.* Albany, N.Y.: Steam Press of C. Van Benthuysen, 1862.

Rumph, J. D. "Thoughts on Malaria, and the Causes Generally of Fever." *Charleston Medical Journal and Review* 9 (July 1854): 439–46.

Sanders, Mary Elizabeth, ed. "Letters of a Confederate Soldier, 1862–1863." *Louisiana Historical Quarterly* 29 (October 1946): 1229–40.

Sanitary Report of the Army of Arkansas for the Last Quarter of 1863. U.S. Surgeon General's Office, History of Medicine Division, National Library of Medicine, Bethesda, Md.

Satchwell, S. S. "Address of Dr. S. S. Satchwell, on Malaria." *Transactions of the Medical Society of the State of North Carolina.* Wilmington: T. Loring, 1852.

Scott, John. *Story of the Thirty-second Iowa Infantry Volunteers.* Nevada, Iowa: John Scott, 1896.

Sherman, William T. *Memoirs of Gen. W. T. Sherman, Written by Himself,* Vol. 1. New York: Charles L. Webster & Co., 1891.

Simons, Thomas Y. *An Essay on the Yellow Fever, as It Has Occurred in Charleston, Including Its Origin and Progress up to the Present Time.* Charleston, S.C.: Steam Power–Press of Walker and James, 1851.

Smiley, Thomas T. "The Yellow Fever at Port Royal, S.C." *Boston Medical and Surgical Journal* 67, January 8, 1863, 449–68.

Smith, Ashbel. "On the Climate, Etc., of a Portion of Texas." *Southern Medical Reports* 2 (1850): 453–59.

Southern Historical Society Papers. 52 vols. Richmond, Va.: Southern Historical Society, 1876–1959.

Sperry, A. F. *History of the 33d Iowa Infantry Volunteer Regiment, 1863–6.* Des Moines: Mills & Co., 1866.

Stampp, Kenneth M., ed. *Records of Ante-bellum Southern Plantations from the Revolution through the Civil War.* Microfilm. Frederick, Md.: University Publications of America, 1985–2000.

Stillwell, Leander. *The Story of a Common Soldier of Army Life in the Civil War, 1861–1865.* Erie, Kan.: Franklin Hudson Publishing Co., 1920.

Storrs, Richard S. *Terrors of the Pestilence: A Sermon, Preached in the Church of the Pilgrims, Brooklyn, N.Y., on Occasion of a Collection in Aid of the Sufferers at Norfolk, Va., September 30th, 1855.* New York: John A. Gray, 1855.

Taylor, Thomas E. *Running the Blockade: A Personal Narrative of Adventures, Risks, and Escapes during the American Civil War.* 1897. Reprint. Annapolis, Md.: Naval Institute Press, 1995.

Thompson, Edwin Porter. *History of the First Kentucky Brigade.* Cincinnati: Caxton Publishing House, 1868.

Thompson, Samuel. "Malaria and Its Relation to the Existence and Character of Disease." *North-Western Medical and Surgical Journal* 3 (May 1850): 19–43.

Thorndike, Rachel Sherman, ed. *The Sherman Letters: Correspondence between General Sherman and Senator Sherman from 1837 to 1891.* New York: Da Capo Press, 1969.

Thorpe, Sheldon B. *The History of the Fifteenth Connecticut Volunteers in the War for the Defense of the Union, 1861–1865.* New Haven, Conn.: Price, Lee & Adkins, Co., 1893.

Todd, William. *The Seventy-ninth Highlanders: New York Volunteers in the War of the Rebellion, 1861–1865.* Albany: Brandow, Barton, & Co., 1886.

Towne, Laura. *Letters and Diary of Laura M. Towne. Written from the Sea Islands of South Carolina, 1862–1884.* Edited by Rupert Sargent Holland. New York: Negro Universities Press, 1969.

Urquhart, Kenneth Trist, ed. *Vicksburg, Southern City under Siege: William Lovelace Foster's Letter Describing the Defense and Surrender of the Confederate Fortress on the Mississippi.* New Orleans: Historic New Orleans Collection, 1980. Housed with United States Sanitary Commission Records, Library of Congress.

Vaché, Alexander F. *Letters on Yellow Fever, Cholera, and Quarantine; Addressed to the Legislature of the State of New York.* New York: McSpedon & Baker, 1852.

Vandiver, Frank E. *Confederate Blockade Running through Bermuda, 1861–1865: Letters and Cargo Manifests.* Austin: University of Texas Press, 1947.

Walker, T. H. "Camp Diarrhœa." *Chicago Medical Journal* 5 (August 1862): 478–80.

Walkley, Stephen. *History of the Seventh Connecticut Volunteer Infantry, Hawley's Brigade, Terry's Division, Tenth Army Corps, 1861–1865.* Hartford, Conn., 1905.

War Department Collection of Confederate Records. Record Group 109, National Archives. Washington, D.C.

The War of the Rebellion: A Compilation of the Official Records of the Union and Confederate Armies. 128 vols. Washington, D.C.: Government Printing Office, 1880–1901.

War Papers Read before the Commandery of the State of Maine, Military Order of the Loyal Legion of the United States, Vol. 4. Portland, Maine: Lefavor-Tower Co., 1915.

War Papers Read before the Commandery of the State of Wisconsin, Military Order of the Loyal Legion of the United States. Milwaukee: Burdick, Armitage & Allen, 1896.

Warren, Edward. "Observations on Miasmatic Diseases." *American Medical Monthly* 2 (October 1854): 241–53.

Webb, William. "On the So-Called Malarious Diseases. A Paper Read before the St. Louis Medical Society." *St. Louis Medical and Surgical Journal* 12 (December 1854): 481–85.

"The Week." *American Medical Times* 7, August 8, 1863, 65–66.

Weld, Stanley B. "A Connecticut Surgeon in the Civil War: The Reminiscences of Dr. Nathan Mayer." *Journal of the History of Medicine and Allied Sciences* 19 (July 1964): 272–86.

Wheaton, James W., comp. *Surgeon on Horseback: The Missouri and Arkansas Journal and Letters of Dr. Charles Brackett of Rochester, Indiana, 1861–1863.* Carmel: Guild Press of Indiana, 1998.

Whitcomb, Caroline E. *History of the Second Massachusetts Battery (Nim's Battery) of Light Artillery, 1861–1865.* Concord, N.H.: Rumford Press, 1912.

Williams, G. Mott. "Letters of General Thomas Williams, 1862." *American Historical Review* 14 (January 1909): 304–28.

Willson, Geo. B. "Army Ambulances—Cases in the Hospital of Richardson's Brigade." *Boston Medical and Surgical Journal* 65 (January 1862): 542–44.

Winschel, Terry J., ed. *The Civil War Diary of a Common Soldier: William Wiley of the 77th Illinois Infantry.* Baton Rouge: Louisiana State University Press, 2001.

Woodward, Joseph Janvier. *Outlines of the Chief Camp Diseases of the United States Armies as Observed during the Present War.* 1863. Reprint. San Francisco: Norman Publishing, 1992.

Worsham, W. J. *The Old Nineteenth Tennessee Regiment, C.S.A., June 1861. April, 1865.* Knoxville: Press of Paragon Printing Co., 1902.

Wragg, W. T. "Report on the Epidemic of Yellow Fever Which Prevailed at Wilmington, N.C., in the Fall of 1862." *New York Medical Journal* 9 (August 1869): 478–96.

Wright, Henry H. *A History of the Sixth Iowa Infantry.* Iowa City: State Historical Society of Iowa, 1923.

Younger, Edward. *Inside the Confederate Government: The Diary of Robert Garlick Hill Kean, Head of the Bureau of War.* New York: Oxford University Press, 1957.

Periodicals

Austin State Gazette
Bellville Countryman
Bermuda Royal Gazette
Charleston Daily Courier
Charleston Mercury
Galveston Weekly News
Harper's Weekly
Houston Telegraph
Houston Tri-Weekly Telegraph
Natchitoches Union
New Orleans Daily Picayune
New York Herald
New York Times
San Antonio Semi-Weekly News
Savannah Daily Morning News
Wilmington Journal

SECONDARY SOURCES

Ackerknecht, Erwin H. *Malaria in the Upper Mississippi Valley, 1760–1900.* Baltimore: Johns Hopkins Press, 1945.

Adams, George Worthington. "Confederate Medicine." *Journal of Southern History* 6 (May 1940): 151–66.

———. *Doctors in Blue: The Medical History of the Union Army in the Civil War.* Baton Rouge: Louisiana State University Press, 1952.

Anderson, Bern. "The Naval Strategy of the Civil War." *Military Affairs* 26 (Spring 1962): 11–21.

Axelrod, Alan. *The War between the Spies: A History of Espionage during the American Civil War.* New York: Atlantic Monthly Press, 1992.

Baird, Nancy Disher. *Luke Pryor Blackburn: Physician, Governor, Reformer.* Lexington: University Press of Kentucky, 1979.

Ballard, Michael B. *Vicksburg: The Campaign That Opened the Mississippi*. Chapel Hill: University of North Carolina Press, 2004.

Bastian, David F. *Grant's Canal: The Union Attempt to Bypass Vicksburg*. Shippensburg, Pa.: Burd Street Press, 1995.

Beringer, Richard E., Herman Hattaway, Archer Jones, and William N. Still Jr. *Why the South Lost the Civil War*. Athens: University of Georgia Press, 1986.

Blessington, Joseph Palmer. *The Campaigns of Walker's Texas Division*. Austin, Tex.: State House Press, 1994.

Blanton, Wyndham B. *Medicine in Virginia in the Nineteenth Century*. Richmond: Garrett & Massie, 1933.

Block, W. T. "Yellow Fever Plagued Area during 1860s." *Beaumont Enterprise*, August 7, 1999.

Bollet, Alfred Jay. *Civil War Medicine: Challenges and Triumphs*. Tucson: Galen Press, 2002.

———. *Plagues & Poxes: The Impact of Human History on Epidemic Disease*. 2nd ed. New York: Demos, 2004.

Boyd, Mark F. "An Historical Sketch of the Prevalence of Malaria in North America." *American Journal of Tropical Medicine* 21 (March 1941): 223–44.

Breeden, James O. "A Medical History of the Later Stages of the Atlanta Campaign." *Journal of Southern History* 35 (February 1969): 31–59.

Burgess, N.R.H., and G. O. Cowan. *Atlas of Medical Entomology*. London: Chapman & Hall Medical, 1993.

Butcher, Carol A. "Malaria: A Parasitic Disease." *American Association of Occupational Health Nurses Journal* 52 (July 2004): 302–9.

Capers, Gerald M., Jr. "Confederates and Yankees in Occupied New Orleans, 1862–1865." *Journal of Southern History* 30 (November 1964): 405–26.

Caplan, Arthur L., H. Tristram Engelhardt Jr., and James L. McCartney, eds. *Concepts of Health and Disease: Interdisciplinary Perspectives*. Reading, Mass.: Addison-Wesley, 1981.

Carrier, John. "Medicine in Texas: The Struggle with Yellow Fever, 1839–1903." *Texas Medicine* 82 (November 1986): 62–65.

Carrigan, Jo Ann. *The Saffron Scourge: A History of Yellow Fever in Louisiana, 1796–1905*. Lafayette: University of Southwestern Louisiana, 1994.

———. "Yankees versus Yellow Jack in New Orleans, 1862–1866." *Civil War History* 9 (September 1963): 248–61.

Carter, H. R. *Yellow Fever: Its Epidemiology, Prevention, and Control*. Washington, D.C.: Government Printing Office, 1914.

Catton, Bruce. *Grant Moves South*. Boston: Little, Brown, 1960.

Christophers, Rickard. *Aedes Aegypti: The Yellow Fever Mosquito*. Cambridge: Cambridge University Press, 1960.

Colbourne, Michael. *Malaria in Africa*. London: Oxford University Press, 1966.

Cooke, Michael. "The Health of the Union Military in the District of Columbia, 1861–1865." *Military Affairs* 48 (October 1984): 194–99.

Cotham, Edward T., Jr. *Sabine Pass: The Confederacy's Thermopylae*. Austin: University of Texas Press, 2004.

Cumberland, Charles C. "The Confederate Loss and Recapture of Galveston, 1862–1863." *Southwestern Historical Quarterly* 51 (October 1947): 109–19.

Cunningham, H. H. *Doctors in Gray: The Confederate Medical Service*. Baton Rouge: Louisiana State University Press, 1958.

———. "The Medical Services and Hospitals of the Southern Confederacy." Ph.D. diss., University of North Carolina, 1952.

Darsie, Richard F., Jr., and Ronald A. Ward. *Identification and Geographical Distribution of the Mosquitoes of North America, North of Mexico*. Gainesville: University Press of Florida, 2005.

Davis, Stephen. *Atlanta Will Fall: Sherman, Joe Johnston, and the Yankee Heavy Battalions*. Wilmington, Del.: Scholarly Resources, 2001.

Denney, Robert E. *Civil War Medicine: Care & Comfort of the Wounded*. Foreword by Edwin C. Bearss. New York: Sterling Publishing, 1994.

Diamond, William. "Imports of the Confederate Government from Europe and Mexico." *Journal of Southern History* 6 (November 1940): 470–503.

Dirckx, John H., ed. *Stedman's Concise Medical Dictionary for the Health Professions*. 4th ed. Philadelphia: Lippincott, Williams, & Wilkins, 2001.

Duffy, John. "Medical Practices in the Ante Bellum South." *Journal of Southern History* 25 (February 1959): 53–72.

———. "Yellow Fever in the Continental United States during the Nineteenth Century." *Bulletin of the New York Academy of Medicine* 44 (June 1968): 687–701.

Eicher, David J. *The Longest Night: A Military History of the Civil War*. New York: Simon & Schuster, 2001.

Farnham, Thomas J., and Francis P. King. "'The March of the Destroyer': The New Bern Yellow Fever Epidemic of 1864." *North Carolina Historical Review* 73 (October 1996): 435–83.

Fenn, Elizabeth. *Pox Americana: The Great Smallpox Epidemic of 1775–82*. New York: Hill & Wang, 2001.

Flannery, Michael A. *Civil War Pharmacy: A History of the Drugs, Drug Supply and Provision, and Therapeutics for the Union and Confederacy*. New York: Pharmaceutical Products Press, 2004.

Foltz, Charles S. *Surgeon of the Seas: The Adventurous Life of Surgeon General Jonathan M. Foltz in the Days of Wooden Ships*. Indianapolis: Bobbs-Merrill, 1931.

Foote, Shelby. *The Civil War*. 3 vols. New York: Vintage Books, 1986.

Franke, Norman H. "Rx Prices in the Confederacy." *Journal of the American Pharmaceutical Association* 12 (December 1961): 773–74.

Freemon, Frank R. *Gangrene and Glory: Medical Care during the American Civil War.* Urbana: University of Illinois Press, 1998.

———. "Medical Care at the Siege of Vicksburg, 1863." *Bulletin of the New York Academy of Medicine* 67 (September–October 1991): 429–38.

———. "The Medical Challenge of Military Operations in the Mississippi Valley during the American Civil War." *Military Medicine* 157 (September 1992): 494–97.

Goldfield, David R. "The Business of Health Planning: Disease Prevention in the Old South." *Journal of Southern History* 42 (November 1976): 557–70.

Hallock, Judith Lee. "'Lethal and Debilitating': The Southern Disease Environment as a Factor in Confederate Defeat." *Journal of Confederate History* 7 (1991): 51–61.

Hasegawa, Guy R. "Pharmacy in the American Civil War." *American Journal of Health-System Pharmacy* 57 (March 2000): 475–89.

Hayes, Charles W. *History of the Island and the City of Galveston.* 2 vols. Austin: Jenkins Garrett Press, 1974.

Hoch, J. Hampton. "Through the Blockade." *Journal of the American Pharmaceutical Association* 12 (December 1961): 769–70.

Holyman, Robert S. "Ben Butler in the Civil War." *New England Quarterly* 30 (September 1957): 330–45.

Horsfall, William R. *Medical Entomology: Arthropods and Human Disease.* New York: Ronald Press Co., 1962.

Humphreys, Margaret. *Malaria: Poverty, Race, and Public Health in the United States.* Baltimore: Johns Hopkins University Press, 2001.

———. *Yellow Fever and the South.* Baltimore: Johns Hopkins University Press, 1992.

Jarecka, Louise Llewellyn. "Virginia Moon, Unreconstructed Rebel." *Delphian Quarterly* 30 (January 1947): 17–21.

Joiner, Gary Dillard. *One Damn Blunder from Beginning to End: The Red River Campaign of 1864.* Wilmington, Del.: Scholarly Resources, 2003.

King, Joseph Edward. "Shoulder Straps for Aesculapius: The Vicksburg Campaign, 1863." *Military Surgeon* 114 (March 1954): 216–26.

Kiple, Kenneth F. "Black Yellow Fever Immunities, Innate and Acquired, as Revealed in the American South." *Social Science History* 1 (Summer 1977): 419–36.

———. "A Survey of Recent Literature on the Biological Past of the Black." *Social Science History* 10 (Winter 1986): 343–67.

Koeniger, A. Cash. "Climate and Southern Distinctiveness." *Journal of Southern History* 54 (February 1988): 21–44.

Kupperman, Karen Ordahl. "Fear of Hot Climates in the Anglo-American Colonial Experience." *William and Mary Quarterly* 41 (April 1984): 213–40.

Leonard, Elizabeth D. *All the Daring of the Soldier: Women of the Civil War Armies.* New York: W. W. Norton, 1999.

Mandell, Gerald L., John E. Bennett, and Raphael Dolin, eds. *Principles and Practice of Infectious Diseases.* 5th ed. Philadelphia: Churchill Livingstone, 2000.

Massey, Mary Elizabeth. *Bonnet Brigades.* New York: Alfred A. Knopf, 1966.

———. *Ersatz in the Confederacy: Shortages and Substitutes on the Southern Homefront.* Columbia: University of South Carolina Press, 1993.

McComb, David G. *Galveston: A History.* Austin: University of Texas Press, 1986.

McCormmach, Russell. "Ormsby MacKnight Mitchel's 'Sidereal Messenger,' 1846–1848." *Proceedings of the American Philosophical Society* 110, February 18, 1966, 35–47.

McMurry, Richard M. "Marse Robert and the Fevers: A Note on the General as Strategist and on Medical Ideas as a Factor in Civil War Decision Making." *Civil War History* 35 (September 1989): 197–207.

McNeill, William H. *Plagues and Peoples.* Garden City, N.Y.: Anchor Books, 1976.

McPherson, James M. *Battle Cry of Freedom: The Civil War Era.* New York: Ballantine Books, 1988.

———. *Crossroads of Freedom: Antietam.* New York: Oxford University Press, 2002.

Miller, Edward S. "A Long War and a Sickly Season." *Blue & Gray Magazine* 9 (June 1992): 40–44.

Mitchel, F. A. *Ormsby MacKnight Mitchel, Astronomer and General: A Biographical Narrative.* Boston: Riverside Press, 1887.

Mitchell, Reid. *Civil War Soldiers.* New York: Penguin Books, 1988.

Moulton, Forest Ray, ed. *A Symposium on Human Malaria with Special Reference to North America and the Caribbean Region.* Washington, D.C.: American Association for the Advancement of Science, 1941.

Ockenhouse, Christian F., Alan Magill, Dale Smith, and Wil Milhous. "History of U.S. Military Contributions to the Study of Malaria." *Military Medicine* 170 (April 2005): 12–16.

Parks, Joseph H. "A Confederate Trade Center under Federal Occupation: Memphis, 1862 to 1865." *Journal of Southern History* 7 (August 1941): 289–314.

Parment, Sharon. "Malaria." *Journal of the American Medical Association* 291, June 2, 2004, 2664.

Patterson, Gordon. *The Mosquito Wars: A History of Mosquito Control in Florida.* Gainesville: University Press of Florida, 2004.

Patterson, K. David. "Yellow Fever Epidemics and Mortality in the United States, 1693–1905." *Social Science and Medicine* 34 (1992): 855–65.

Pearce, George F. *Pensacola during the Civil War*. Gainesville: University Press of Florida, 2000.

Rivers, Thomas M. *Viral and Rickettsial Infections of Man*. 2nd ed. Philadelphia: J. B. Lippincott Co., 1952.

Robertson, James I., Jr. *Soldiers Blue and Gray*. Columbia: University of South Carolina Press, 1988.

Rose, Willie Lee. *Rehearsal for Reconstruction: The Port Royal Experiment*. London: Oxford University Press, 1964.

Ross, Michael A. "Justice Miller's Reconstruction: The Slaughter-House Cases, Health Codes, and Civil Rights in New Orleans, 1861–1873." *Journal of Southern History* 64 (November 1998): 649–76.

Rutman, Darrett B., and Anita H. Rutman. "Of Agues and Fevers: Malaria in the Early Chesapeake." *William and Mary Quarterly* 33 (January 1976): 31–60.

Savitt, Todd L., and James Harvey Young, eds. *Disease and Distinctiveness in the American South*. Knoxville: University of Tennessee Press, 1988.

Schmidt, Lewis G. *A Civil War History of the 47th Regiment of Pennsylvania Veteran Volunteers: The Wrong Place at the Wrong Time*. Allentown, Pa.: Lewis G. Schmidt, 1986.

———. *The Civil War in Florida*. 4 vols. Allentown, Pa.: Lewis G. Schmidt, 1992.

Simkins, Francis B., and James W. Patton. "The Work of Southern Women among the Sick and Wounded of the Confederate Armies." *Journal of Southern History* 1 (November 1935): 475–96.

Smith, Dale C. "The Rise and Fall of Typhomalarial Fever: I. Origins." *Journal of the History of Medicine and Allied Sciences* 37 (April 1982): 182–220.

Spielman, Andrew, and Michael D'Antonio. *Mosquito: A Natural History of Our Most Persistent and Deadly Foe*. New York: Hyperion, 2001.

Steers, Edward, Jr. "A Rebel Plot and Germ Warfare." *Washington Times*, November 10, 2001.

———. "Risking the Wrath of God." *North & South* 3 (September 2000): 59–70.

Steiner, Paul. *Disease in the Civil War: Natural Biological Warfare in 1861–1865*. Springfield, Ill.: Charles C. Thomas, 1968.

———. *Medical History of a Civil War Regiment: Disease in the Sixty-fifth United States Colored Infantry*. N.p.: Institute of Civil War Studies, 1977.

Strode, Hudson. *Jefferson Davis: American Patriot, 1808–1861*. New York: Harcourt, Brace, 1955.

Tidwell, William A., James O. Hall, and David Winfred Gaddy. *Come Retribution: The Confederate Secret Service and the Assassination of Lincoln*. Jackson: University Press of Mississippi, 1988.

Underwood, Anne. "Tracking Disease." *Newsweek* 146, November 14, 2005, 46–48.

Underwood, Rodman L. *Waters of Discord: The Union Blockade of Texas during the Civil War.* London: McFarland & Co., 2003.

Wahlgren, Mats, and Peter Perlmann, eds. *Malaria: Molecular and Clinical Aspects.* Amsterdam: Overseas Publishing Association, 1999.

Waring, Joseph I. *A History of Medicine in South Carolina, 1825–1900.* Charleston: South Carolina Medical Association, 1967.

Warrell, David A., and Hebert M. Gilles, eds. *Essential Malariology.* 4th ed. London: Arnold, 2002.

Warren, Christopher. "Northern Chills, Southern Fevers: Race-Specific Mortality in American Cities, 1730–1900." *Journal of Southern History* 63 (February 1997): 23–56.

Watson, Alan D. *A History of New Bern and Craven County.* New Bern: Tryon Palace Commission, 1987.

Watts, Sheldon. *Epidemics and History: Disease, Power, and Imperialism.* New Haven: Yale University Press, 1997.

Wiley, Bell I. *The Life of Billy Yank.* New York: Bobbs-Merrill, 1951.

———. *The Life of Johnny Reb.* New York: Bobbs-Merrill, 1943.

Wood, Peter H. *Black Majority: Negroes in Colonial South Carolina from 1670 through the Stono Rebellion.* New York: W. W. Norton, 1974.

Wood, Richard Everett. "Port Town at War: Wilmington, North Carolina 1860–1865." Ph.D. diss., Florida State University, 1976.

Wooster, Ralph A., ed. *Lonestar Blue and Gray: Essays on Texas in the Civil War.* Austin: Texas State Historical Association, 1995.

"Yellow Fever Vaccine." *World Health Organization Weekly Epidemiological Record* 78 (2003): 349–59.

Young, Kevin R. *To the Tyrants Never Yield.* Plano: Wordware Publishing, 1992.

INDEX